CATALYSIS IN COAL CONVERSION

CATALYSIS IN COAL CONVERSION

JAMES A. CUSUMANO
RALPH A. DALLA BETTA
RICARDO B. LEVY

Catalytica Associates, Inc.
Santa Clara, California

ACADEMIC PRESS New York San Francisco London 1978

A Subsidiary of Harcourt Brace Jovanovich, Publishers

ACADEMIC PRESS, INC.
111 Fifth Avenue, New York, New York 10003

United Kingdom Edition published by
ACADEMIC PRESS, INC. (LONDON) LTD.
24/28 Oval Road, London NW1 7DX

Library of Congress Cataloging in Publication Data

Cusumano, James A

Catalysis in coal conversion.

Includes bibliographical references and index.
1. Catalysis. 2. Catalysts. 3. Coal liquefaction.
4. Coal gasification. I. Dalla Betta, Ralph A., joint author. II. Levy, Ricardo, joint author. III. Farkas, Adalbert. IV. Title.
TP156.C35C87 662'.6622 77-25620
ISBN 0-12-199935-1

PRINTED IN THE UNITED STATES OF AMERICA

Contents

Preface

One of the major challenges of our time is the solution of the energy problem. The feasibility of the catalytic conversion of coal into liquid and gaseous fuels has been demonstrated, but there is a need for a significant technical breakthrough to make coal conversion an economically attractive process and to permit coal-derived fuels to play an important part in replacing or supplementing our critical oil and gas supplies.

This book marshals the recent advances in catalysis and related disciplines and shows how they might make decisive contributions toward solutions of the catalytic problems of coal conversion.

The book is written primarily for the chemist and chemical engineer engaged in research on some aspect of the conversion of coal into synthetic fuels or of fuel processing. It should also be useful to scientists or technologists concerned with, or interested in, catalysts or catalytic processes in general in view of the wealth of information presented on the preparation, mode of action, and behavior of numerous catalysts, and on the ways their activity, selectivity, and stability can be improved. Teachers of chemistry will find many interesting and instructive examples of practical applications in industrial catalysis for the results of fundamental and theoretical studies.

Part I surveys the advances of catalysis in the past decade or two and covers bimetallic catalysts; the effects of catalyst–support interactions, particle size, and surface morphology; characterization and preparation of catalysts; poisoning and regeneration; and reaction mechanisms relevant to coal conversion.

Part II summarizes the accomplishments of such related disciplines as reactor engineering, inorganic and organometallic chemistry, materials science, and surface science. The topics discussed include catalyst evaluation; novel inorganic, intermetallic, and organometallic compounds of potential catalytic interest; sintering; new concepts of surface layers; and modern spectroscopic methods for the examination of solid surfaces.

Part III, essentially the second half of the book, reviews the nature of the various coal liquefaction processes and their products. It also covers coal gasification and the synthesis of liquid fuels from carbon monoxide–hydrogen mixtures. In each, the catalytic problems are highlighted, and the potential impact of the recent advances in catalysis and related disciplines on solving these problems are critically evaluated.

Adalbert Farkas
Editor

Acknowledgment

This book is based on a report entitled "Scientific Resources Relevant to the Catalytic Problems in the Conversion of Coal" prepared by the authors for the U.S. Energy Research and Development Administration (now U.S. Department of Energy), Fossil Energy, Division of Materials and Exploratory Research.

The diligent and constructive efforts of Dr. Adalbert Farkas in editing this report for publication as the present volume are gratefully acknowledged.

James A. Cusumano
Ralph A. Dalla Betta
Ricardo B. Levy

Chapter 1

Introduction

The conversion of coal to synthetic fuels encompasses a number of catalytic processes and reactions, the most important of which are summarized in Table 1-1. In addition to various current processes of coal conversion to gaseous and liquid fuels, this list includes many catalytic reactions used in the refining of petroleum. Similar reactions are expected to be applicable for upgrading coal liquids. Therefore, the coal conversion industry is most likely to utilize a significant fraction of presently known catalytic materials, including metals, oxides, solid acids, and sulfides.

Paralleling this projected development of catalyst technology for coal conversion are a number of technical challenges familiar to the catalytic chemist and engineer. These include control of selectivity and activity; minimizing catalyst poisoning, sintering, and mechanical degradation; and the development of efficient regeneration procedures and of reactor designs for optimum heat and mass transfer. Because of the complex nature of coal, these challenges are more difficult than those encountered in the development of petroleum processes, but it is expected that the advances in catalysis during the last decade will play a significant role in solving many of the inherent constraints and problems of coal conversion processes. This book provides a perspective of recent advances in catalysis and related disciplines and analyzes their impact on the current status and future development of the technology of synthetic fuels derived from coal.

TABLE 1-1
CATALYTIC REACTIONS IN COAL CONVERSION

Process	General reactions	Specific reactions/products
Direct liquefaction	Hydrogenation	Aromatic liquids
	Cracking	
	Hydrofining	Hydrodesulfurization (HDS)
		Hydrodenitrogenation (HDN)
CO/H_2 synthesis	Fischer–Tropsch	Methane
		Hydrocarbon liquids
		Alcohols
		Chemicals
Water–gas shift		Hydrogen
Direct gasification	Hydrogasification	Methane
	Oxidative gasification	Synthesis gas
Liquids refining and upgrading	Cracking	Hydrogenation
	Reforming	Dehydrogenation
	Hydroforming	Dehydrocyclization
		Isomerization
		Hydrogenolysis
		Hydrodesulfurization (HDS)
		Hydrodenitrogenation (HDN)

The discussions in Part I cover six broad areas: multimetallic catalysts; effects of catalyst-support interactions, particle size, and surface morphology on catalyst properties; catalyst characterization and preparation; poisoning and regeneration; and reaction mechanisms of importance to catalytic coal conversion. While the emphasis in these areas is on work conducted during the last decade, some relevant information concerning earlier work is included. In addition to a review of the main research topics, the general implications of these topics to coal conversion are outlined.

Part II reviews advances in supporting disciplines related to catalysis under four general subjects: reactor engineering and testing procedures; inorganic and organometallic chemistry of catalysts; certain aspects of material science relevant to supports, sintering, and novel materials; and some concepts of surface science relating to metal catalysis and techniques for materials analysis and characterization.

A summary of the most significant developments discussed in Parts I and II, and their impact on coal conversion technology, is presented in Table 1-2.

Part III is devoted to the detailed review of the specific applications of catalysis to a number of important coal conversion processes. These applications include upgrading of coal liquids derived from the Coalcon, COED, H-Coal, and

TABLE 1-2
ADVANCES OF IMPORTANCE TO THE CATALYTIC CONVERSION OF COAL

Subject	Representative areas of impact
Multimetallic catalysis	Upgrading of coal liquids Methanation Fischer–Tropsch synthesis
Catalyst-support interactions	Thermal and chemical stabilization of catalysts and supports
Catalyst characterization	Improvement of coal liquefaction catalysts Upgrading of coal liquids Determination of catalyst *intrinsic* activity
Catalyst preparation	New catalyst formulations Controlled variation in catalyst properties Higher surface area catalysts
Poisoning and regeneration	Effect of sulfur in coal conversion catalysis Prevention of carbon deposition—removal of carbon
Mechanism and surface chemistry	Better understanding of the important steps in coal conversion Identification of rate limiting processes, directions for process improvement
Reactor engineering and catalyst testing	Development of effective catalyst testing procedures Data interpretation
Inorganic chemistry	New catalytic materials and compositions with improved poison tolerance and activity
Materials science	Novel support materials and structures Novel refractory compounds Understanding of sintering phenomena (catalyst deactivation)
Surface science	New characterization techniques

Synthoil processes; the liquefaction of coal and the conversion of solvent refined coal to boiler fuel low in sulfur and nitrogen; catalytic gasification of coal, shift conversion, and methanation; and the synthesis of diesel fuel, LPG, and selected feedstocks from CO and H_2.

The discussions include a brief review of each process and a summary of the major problems and constraints associated with the catalytic steps. Short- and long-range programs are treated separately, as are developments of fundamental nature.

PART I

SURVEY OF ADVANCES IN CATALYSIS

Chapter 2

Multimetallic Catalysts

2.1 GENERAL NATURE OF MULTIMETALLIC CATALYSTS

One of the significant advances in catalysis over the last decade is the development of multimetallic systems for the control of catalytic properties. The impact of this research on technology is already apparent and is likely to increase substantially over the next decade. New multimetallic catalysts with high activity and unprecedented activity maintenance have been commercialized for the catalytic reforming of petroleum naphthas to high-octane gasoline [1–4]. Similar systems are finding utility in other petroleum processes such as isomerization [5], hydrocracking [6], and hydrogenation [7]. In the petrochemicals field, research on bimetallic catalysts has led to improved Pd–Au catalysts for the synthesis of vinyl acetate [8], and to more selective catalysts (e.g., Ag–Au, Cu–Au) for the partial oxidation of olefins [9, 10]. Still other studies have uncovered supported alloys such as Pt–Co [11] which have increased thermal stability and resistance to sintering. Product selectivity is a critical parameter which can now be optimized for many processes by the use of multimetallic systems. For example, in dehydrocyclization [12] and aromatic hydrogenation [13], it is possible to minimize cracking reactions (hydrogenolysis) and maximize the yield of desired products while maintaining very high hydrogenation activity.

In view of the expected role of multimetallic catalysts in present and emerging petroleum and petrochemical technology it was considered desirable to review the advances made in this area, and to highlight the aspects likely to impact on the catalytic conversion of coal.

Although some patents and publications disclose multimetallic catalysts, for sake of simplicity the present discussion will be limited primarily to the bimetallic system. The general nature of bimetallic catalysts, their physical, chemical, and catalytic properties, their preparation and characterization, and their applications in reactions involved in the catalytic conversion of coal will be reviewed.

Bimetallic systems have been of interest to catalytic scientists for some time. Much of the initial work in this area concerns the relationship between catalytic activity and the electronic structure of metals, and dates back to the early concepts proposed by Dowden [14, 15] and Schwab [16]. The approach in that work was the study of catalytic activity as a function of alloy composition, the latter determining the electronic properties of the metal. Alloys of a Group VIII and a Group IB metal (e.g., Ni–Cu) have received particular attention in this regard because it has been commonly suggested that the d electrons of the metal play an important role in determining catalytic activity. For these alloys the IB metal is considered to donate s electrons to the d band of the Group VIII metal and thus, by this model, it is thought possible to control the d electron density. Bimetallic alloys of Group IB with Group VIII metals have been studied by a number of workers for such reactions as the hydrogenation of benzene [17–20] and ethylene [21, 22]. Some of these early concepts, while useful as working hypotheses, are now known to be inadequate for understanding the properties of bimetallic systems. More realistic models, such as the virtual bound state concept [23, 24], provide a better rationale than the simple band picture for many of the physical, chemical, and catalytic properties of bimetallic systems.

The development of new characterization techniques has played a significant role in understanding the properties of bimetallic catalysts [25]. For example, surface analysis techniques such as Auger spectroscopy and selective adsorption have shown that the surface and bulk compositions of a bimetallic catalyst can differ significantly [13, 20]. Such information is critical for the interpretation of catalytic data for bimetallic systems [26].

In general, bimetallic catalysts can be divided into two broad classes, supported and unsupported. The latter has received extensive attention over the years because single crystals, films, and powders of alloys are, in general, easier to study and characterize. More recent efforts, however, have been directed toward supported bimetallic catalysts because of the technological importance of such systems.

Supported bimetallic catalysts can be further subdivided into two types, alloys and clusters. An alloy catalyst consists of two mutually soluble metals having a composition well defined by a bulk phase diagram. The bulk composition of a bimetallic alloy catalyst is usually obtainable by x-ray diffraction methods, since in many cases there is an almost linear relationship between lattice constant and alloy composition (Vergard's law). These measurements can be readily made if the catalyst exists in a sufficiently large crystallite size (>30 Å) to minimize line broadening effects in the x-ray pattern.

Highly dispersed bimetallic alloy catalysts can approach crystallite sizes for which most of the metal atoms are surface atoms. These cannot be studied easily by x-ray diffraction and are called bimetallic clusters [26]. This terminology is preferred to the term alloy because many systems of interest in this category include metallic combinations which do not correspond to known bulk alloys [26]. The existence of such bimetallic clusters for systems with major miscibility limitations in the bulk implies that the degree of metal dispersion has a strong effect on the stability of bimetallic clusters. Consequently, physical or chemical conditions which favor crystallite growth for such systems can cause phase separation of the two constituent metals.

Bimetallic cluster catalysts are usually supported on a high surface area material because it is difficult to maintain highly dispersed metals in the unsupported state without resultant sintering and crystallite growth, especially during use at high temperatures. Another reason for supporting these catalysts is to maximize catalyst utilization, particularly in the case of noble metals for which highest activity per gram of metal is necessary for economic reasons. A detailed description of the preparation of bimetallic catalysts is given elsewhere [25, 27].

2.2 PHYSICAL AND CHEMICAL PROPERTIES

The techniques used to characterize bimetallic catalysts include gas adsorption [28], x-ray diffraction [29], magnetic measurements [30], Auger spectroscopy [31], photoelectron spectroscopy [32], and extended x-ray absorption fine structure analysis, EXAFS [33]. Most of these techniques [25] give an indirect measurement of at least one of three properties of bimetallic catalysts: surface area and composition, crystallite size, and the chemical state of the surface metal atoms.

The effects of temperature on a bimetallic catalyst depend on factors that include initial metal dispersion, nature of the support, and the gaseous environment. Much has been learned over the past decade concerning two such effects: catalyst sintering [34, 35] and metal support interactions [36, 37].

Crystallite growth of highly dispersed bimetallic catalysts can cause phase separation of the two metals and change the surface composition. Phase separation is a potential problem with bimetallic clusters composed of two immiscible metals. For example, ruthenium and copper do not form bulk alloys, yet these metals do form relatively stable bimetallic clusters [26]. If the bimetallic RuCu cluster is sintered at high temperatures, the two metals can separate. However, in the dispersed state there is a significant interaction between the two metals which markedly modifies catalytic properties and also gives rise to an increased thermal stability. This is indicated by the decreased tendency for copper to sinter in the RuCu cluster relative to a supported copper catalyst [26]. Similar effects have been observed for OsCu [26] and PtCo [11]. Bimetallic systems therefore offer a means to synthesize catalysts of improved thermal stability.

Change in surface composition as a result of crystallite growth is predictable from the thermodynamic models for dispersed bimetallic systems [38–40]. This effect is observed even for "miscible" systems such as Ni–Cu [41, 42] and Ni–Au [43], and is therefore not restricted to immiscible combinations like Ru–Cu or Os–Cu. The change in surface composition of bimetallic crystallites simply reflects the thermodynamic driving force for the metal with the lowest surface energy to concentrate at the surface. This enrichment increases with decreasing crystallite size [38–40], even to the smallest of crystallites (e.g., 12 Å) for which surface segregation or enrichment can occur [38], and may have marked effects on catalysis. More recent work has shown that the chemisorption of gases can alter surface composition, and studies are now being directed at measurements of surface composition at the conditions of catalysis [44–45].

Chemical stability in the operating environment is an important criterion which a catalyst must meet in a number of reactions involved in the catalytic conversion of coal. The two major demands on a catalyst in such a typical environment are compatibility with sulfur and stability with respect to oxidative regeneration [25].

With respect to sulfur, it has been shown recently [25] that all metals form sulfides in the environment encountered in the direct liquefaction of coal to boiler fuels where H_2S levels of 1% and higher are common. Alloying or formation of bimetallic clusters is not expected to have a significant effect on the thermodynamic driving force for sulfidation under these conditions [25]. However, at lower levels of sulfur contamination (e.g., 10–1000 ppm), cluster or alloy formation may alter the poisoning effects although there are only few data in the literature to support this contention [46]. The example of the resistance of the very stable intermetallic compound $ZrPt_3$ to sulfidation has been discussed previously [25], and it was shown that the more exothermic the heat of formation of the alloy the greater is its thermodynamic resistance to sulfidation. Similar arguments hold for dispersed systems in which bimetallic interactions may affect the nature of the metal–sulfur interaction. The effects of increased metal dispersion are likely to enhance metal–sulfur interactions unless a metal-support effect is present which is expected to be most important at high dispersions, and could cause increased sulfur resistance (see Chapters 3 and 4).

The problems encountered for bimetallic catalysts in oxidative regeneration are twofold. The first is a common concern for the regeneration of all supported metal catalysts, namely sintering of the metal phase. This occurs even when carbon burnoff is controlled carefully by using low oxygen concentrations. The problems of local hot-spotting and high surface temperatures are commonly encountered [47, 48]. After several regeneration cycles the crystallites grow to a size which corresponds to a decrease in active metal surface area, and therefore to lower activity. However, for bimetallic catalysts an additional factor is important. This involves changes in surface composition and phase segregation or separation with increased crystallite size. For some bimetallic systems (e.g., Pt–Co, Pt–Pd, Ru–Cu) there is actually an increased thermal stability with respect to sintering [11, 26]. There is also the possibility that a high-temperature oxidative environment can generate a mixed oxide system for some bimetallic catalysts, especially for noble metal systems. Cycling between the metallic and oxide state in some instances has been reported to give rise to the formation of supported bimetallic catalysts with a different distribution of surface compositions than the original catalyst [49].

On the basis of studies of bimetallic catalysts, these materials are likely to offer a possibility for diminishing the effects of sulfur as long as these catalysts are used at sulfur levels below 1%. They may also lead to improved thermal stability in cases in which the preparative variables are well understood—an area of increasing research interest.

2.3 CATALYTIC PROPERTIES

In the numerous studies concerning the catalytic properties of bimetallic system [13, 25–27, 50], only in the last several years has there been a significant effort to relate catalytic data to surface properties. This is primarily because variations in alloy surface composition from that of the bulk were not considered in earlier studies. Also, only recently have studies dealt with highly dispersed supported bimetallic systems [26, 49]. From these studies it is clear that catalyst preparation variables can have a marked effect on the nature of the bimetallic catalyst [25, 27].

2.3.1 Activity

Although bimetallic systems can be used to improve catalytic activity, it is more informative to consider selectivity data. Often several reactions can occur for a given process, and it is usually necessary to maximize the yield of only one of the resulting products. For example, in catalytic reforming or hydrocracking branched paraffins and aromatics are preferred products, and it is necessary to minimize the yield of light gases such as CH_4 and C_2H_6. Bimetallic catalysts offer a means of increasing the selectivity to the valuable liquid products for such a process.

In terms of activity there have been a number of studies of both simple test reactions and industrial processes. For example, the combination of Group IB with Group VIII metals is reported to increase hydrogenation activity. Thus, Ni–Cu [13] and Ru–Cu [26] show higher rates for cyclohexane dehydrogenation than do the pure metals. However, some investigators noted that this may not be due to greater intrinsic activity of the bimetallic catalyst but rather to less carbon fouling of the bimetallic surface compared to that of the pure metals [50]. In certain systems involving metals with very high hydrogenolysis activity, both factors may be operative. Similar promotional effects are noted for olefin hydrogenation reactions using Pd–Au [7] and Ni–Cu [51, 52].

Recently, bimetallic catalysts have been discovered which show markedly improved activity for the reactions involved in the catalytic reforming of petroleum naphthas to high-octane gasoline. This has brought about the commercialization of such catalysts, including Pt–Re [3], Pt–Pb [53], Pt_3Cu [54], and a multimetallic system called KX-130 [1]. This increase in activity is clearly shown in a study that compares Pt–Re and KX-130 with a conventional platinum reforming catalyst [1] in the conversion of a Middle East paraffinic naphtha at 150 psig and 930°F to high-octane gasoline (102.5 RON). It was found that the initial activity for Pt–Re is comparable to Pt, but the former catalyst has a marked increase in activity maintenance (see Section 2.3.2). The initial activity for the KX-130 catalyst is at least four times higher than that of traditional platinum catalysts. The activity for these catalysts was measured by the space velocity required to give a constant octane of 102.5. The liquid yields for all three catalysts were comparable.

2.3.2 Activity Maintenance

In commercial operation a catalyst can undergo poisoning or fouling (see Chapter 6). For hydrocarbon reactions, this is usually due to poisoning of metal sites by substances such as sulfur or to coke accumulation. For some reactions bimetallic systems have overcome these problems by their increased resistance to sulfur poisoning and coke fouling or decreased coke yields.

An example of the effect of alloying on activity maintenance is provided by the Pt–Re and KX-130 catalysts which show significant resistance to deactivation by coke, apparently due to the decreased deposition of coke [1–3, 50]. After 300 h on stream, the conventional platinum catalyst lost 73% of its initial activity, Pt–Re decreased by 24%, and KX-130 showed no measurable decrease in activity. This increased activity maintenance may occur for two possible reasons. The first is the inhibition of coke accumulation by the polymerization of acetylenic residues, the formation of which is thought to require at least three contiguous metal sites [50]. Contiguous sites can be disrupted by alloying an active metal with an inert metal (e.g., Pt–Au, Pt–Sn, or Pt–Cu), resulting in the dilution of the primary active centers and in the inhibition of coke generation. A second reason is a high but proper balance of hydrogenation and hydrogenolysis activity. The latter promotes cracking of surface residues, and the former facile hydrogenation of the cracked species, and therefore inhibits polymerization of carbon species to coke. Both of these models require a bimetallic catalyst with a proper balance for these key functions.

The ratio of hydrogenation to hydrogenolysis activity of a catalyst can be controlled by the appropriate combination of metals and method of catalyst preparation (see Section 2.3.3) and it has been found [25–27, 55] that generally the high hydrogenolysis to hydrogenation ratios of Group VIII metals (e.g., Ru, Rh, Ir, Ni) can be continuously adjusted by the addition of Group IB metals (Cu, Ag, or Au). This is possible because the IB metals selectively inhibit hydrogenolysis activity without affecting hydrogenation activity adversely. This can have a significant effect on the equilibrium levels of coke deposited on the catalyst as well as product selectivity.

2.3.3 Selectivity

Selectivity is a ratio of activities or rate constants for two reactions occurring simultaneously or consecutively and is a measure of the yield to a given product. Only over the last several years have catalytic studies involving bimetallic systems been directly concerned with selectivity while the majority of prior studies dealt almost solely with catalytic activity [13, 25–27].

In studying Ni–Cu alloys, one of the most widely investigated bimetallic systems, Sachtler and co-workers found a marked enhancement in the selectivity of Ni for C—H scission by alloying it with Cu [56]. This effect was due to the suppression of carbon–carbon bond hydrogenolysis. Similar effects were observed by Sinfelt and co-workers for ethane hydrogenolysis and for cyclohexane dehydrogenation [13] on a series of Ni–Cu alloys. These alloys were characterized by adsorption, x-ray diffraction, and magnetic measurements. The addition of copper to nickel was found to have markedly different effects for the two reactions. The catalytic activity for ethane hydrogenolysis decreased by several orders of magnitude with increasing copper concentration. On the other hand, the cyclohexane dehydrogenation activity increased initially with the first small amounts of copper, and then remained insensitive to alloy composition over a wide range, finally decreasing sharply at very high Cu–Ni atom ratios.

These data can be rationalized on the basis of the rate-limiting steps for the two reactions. For ethane hydrogenolysis, an increased concentration of Ni–Cu pairs on the surface decreases the number of diadsorbed ethane intermediates since hydrocarbons adsorb only weakly on copper. If carbon–carbon bond scission is the rate-limiting step for this reaction, the decrease in reactivity of nickel with the addition of copper can be understood. For cyclohexane dehydrogenation, the rate-limiting step is assumed to be desorption of benzene. In this case, copper is thought to decrease the binding energy of benzene and, therefore, the rate increases. The relative insensitivity of dehydrogenation reaction rate over Ni–Cu alloys of a considerable range of Cu concentrations is attributed to the relatively constant surface composition in this range as determined by H_2 chemisorption. A similar interpretation is given by Burton and Hyman using a model for segregation of alloy surfaces [57].

Some selectivity effects have also been observed for supported bimetallic cluster catalysts [26, 49]. In one case [26], a study of silica-supported Ru–Cu and Os–Cu catalysts for ethane hydrogenolysis and cyclohexane dehydrogenation showed that, in the highly dispersed state, copper interacts strongly with ruthenium and osmium, and decreases the hydrogenolysis activity of these metals, even though copper is virtually immiscible with these metals in the bulk.

More recent work [58] with unsupported Ru–Cu catalysts has shown the same effects in spite of the miscibility problem at larger crystallite sizes. The presence of copper decreases the capacity of ruthenium for hydrogen chemisorption, and suppresses markedly ethane hydrogenolysis. The ethane hydrogenolysis activity of a Ru–Cu catalyst is strikingly related to its capacity for strong hydrogen chemisorption. The interaction between copper and ruthenium occurs at the surface, and is similar to that which would exist if copper were to ''chemisorb'' on ruthenium. The state of dispersion for Ru–Cu catalysts has a major influence on the effect of the copper. The atomic ratio of copper to ruthenium required for a given degree of coverage of the surface by copper increases with increasing dispersion, as is clearly reflected by the results for hydrogen chemisorption and ethane hydrogenolysis.

Other examples of selectivity effects using bimetallic catalysts have been described by Ponec and co-workers for *n*-hexane cyclization over Pt–Au alloys [59] and by Ponec and Sachtler for isomerization over Ni–Cu alloys [60]. The reason for this enhancement in selectivity is ascribed to an increase in the number of isolated metal atoms (e.g., Pt in an Au matrix), whereby reactions and products requiring single sites are favored and products of pathways requiring several contiguous metal sites are suppressed. A further feature of dilute active sites relevant to activity maintenance for hydrocarbon reactions is a decrease in self-poisoning by coke that is thought to form by polymerization of acetylenic residues [50] on larger contiguous arrays of sites [61].

2.4 APPLICATIONS TO THE CONVERSION OF COAL

Table 2-1 lists a number of examples of metal-catalyzed reactions involved in coal conversion.

In discussing any of the reactions listed in this table it should be borne in mind that metallic catalysts in general are not expected to survive as metals in a coal

TABLE 2-1
EXAMPLES OF METAL-CATALYZED REACTIONS

Reaction	Typical metal catalysts	High-activity metals
Hydrogenation		
Aromatics	Group VIII, Mo, W, Re	Pt, Rh, Pd
CO	Group VIII, Cu	Ru, Fe, Ni, Co
CO_2	Co, Fe, Ni, Ru	Ru, Ni
Dehydrogenation	Group VIII	Pt, Pd
Dehydrocyclization	Pt, Pd, Ir, Rh, Ru	Pt, Ir
Hydrogenolysis		
C—C bonds	Group VIII, Re, W, Mo	Ru, Ir, Ni
C—N bonds	Group VIII, Re, W, Mo, Cu	Ni, Pt, Pd
C—O bonds	Group VIII, Re, W, Mo, Cu, Ag	Pt, Pd
C—S bonds	Group VIII, Re, W, Mo, Cu, Ag	Pt, Pd, Rh
Isomerization		
Double bond shift	Group VIII	Pd, Pt
Skeletal	Pt, Ir, Pd, Au	Pt
Water–gas shift	Group VIII, Cu, Ag	Pt, Rh, Fe

liquefaction environment at sulfur levels exceeding 1%. It has been shown [25] that all metals can form bulk sulfides under these conditions, and therefore the true catalyst would be a mixed sulfide and not a bimetallic system. This means that one cannot take advantage of the catalytic properties of bimetallic catalysts, which inevitably depend upon the metallic properties of these systems, for the direct liquefaction of coal to boiler fuel. However, it appears [25] that the concepts uncovered in the evolution of many new bimetallic catalysts can find use in preparing supported highly dispersed mixed sulfides and oxysulfides.

2.4.1 Hydrogenation–Dehydrogenation

The primary applications for bimetallic catalysts in this category fall into two areas: upgrading or refining of processed low-sulfur coal liquids to refined fuels, and CO/H_2 synthesis reactions. The refining processes of interest include hydrocracking, naphtha reforming, and skeletal isomerization. Each process requires a delicate balance between hydrogenation activity and support acidity to minimize gas make and coke formation, and thereby to maximize the yield of liquid products. One cannot draw directly on data for processing petroleum feedstocks which usually have characteristics substantially different from those for coal liquids. However, many of the basic concepts, procedures, and preparative techniques which have been developed for bimetallic systems, and which have led to the discovery of new catalysts with low hydrogenolysis activity, low coke yields, and increased thermal stability, should find application in coal upgrading processes. Studies are needed of supported bimetallic ctalysts with variable hydrogenation to acidity ratios for both model reactions and real feeds. Hydrocracking is perhaps most important in this respect. Work in this area should include studies of catalyst regeneration by oxidative procedures and of the interaction of H_2S with multimetallic catalysts. The apparent increased sulfur tolerance of bimetallics [46] at sulfur levels below 0.1% needs verification and development.

In the CO/H_2 synthesis area (Fischer–Tropsch synthesis), primary problems have been product selectivity, sulfur sensitivity, and deactivation by sintering and carbide formation. Recent work [62, 63] has shown that the binding energy of H_2 and CO to the metal surface can have a strong effect on the CO/H_2 synthesis reaction. Since alloy or cluster formation can affect this energy [64], one would expect improved product selectivity on alloy or cluster catalysts. Decreased sulfur sensitivity and increased stability with respect to sintering are also expected for bimetallic systems.

2.4.2 Hydrogenolysis

As the hydrogenolysis of C—C bonds can be controlled to minimize gas make or coke formation by the use of bimetallic catalysts, such catalysts are likely to find applications in both upgrading coal liquids and in CO/H_2 synthesis. For the latter, decreased hydrogenolysis activity for bimetallic Ni, Ru, or Co catalysts may lead to lower methane yields, a shift in product distribution to liquids, and lower coke formation.

The hydrogenolysis of C—N bonds is an area where more work needs to be done. If the rate-limiting step of the reaction is the C—N scission, it would be of interest to study bimetallic systems which optimize the interaction of carbon and nitrogen with the metallic surface.

Carbon–oxygen hydrogenolysis is a very facile reaction, almost unavoidable in coal processing and a consumer of hydrogen. The applications of bimetallic systems in this area are not clear at this time.

In the hydrogenolysis of C—S bonds (hydrodesulfurization, HDS), a primary reaction of coal processing, oxides, sulfides, oxysulfides, and other nonmetallic materials are likely to be more important than metals which are probably sulfided at sufficiently high sulfur levels.

2.4.3 Water–Gas Shift Reaction

The water–gas shift reaction is usually catalyzed by complex oxide systems such as $ZnO \cdot Cr_2O_3$ and $ZnO \cdot CuO$ [65]. The reaction is also catalyzed by metals; however, in some instances zero-valent catalysts can be severely deactivated because of the facile catalysis of the Boudouard reaction:

$$2CO \rightleftharpoons C + CO_2$$

which can cause substantial carbon deposition on the catalyst. This is very evident with iron oxide catalysts if the H_2O/CO ratio is sufficiently low and the oxide is reduced to metallic iron. Similar effects are expected for other metals. Therefore metallic and multimetallic catalysts would require an additive to minimize carbon deposition by the Boudouard reaction and to prevent deactivation. A positive aspect of the use of multimetallic catalysts for the water–gas shift reaction would be a way for altering the surface bond strengths of CO and H_2, and thereby controlling the rate-limiting step [66].

2.5 SUMMARY

Bimetallic catalysts represent an important chapter in the development of heterogeneous catalysis. Studies with these materials have played a major role in increasing our understanding of the electronic and geometric factors in catalysis by metals. The necessity for differentiating between surface and bulk compositions in such work cannot be overemphasized.

Significant progress has been made in the preparation of bimetallic catalysts and their characterization by improved techniques for surface analysis, including selective chemisorption, Auger spectroscopy, and photoelectron spectroscopy.

Bimetallic catalysts have potential applications for a number of coal processing steps which do not involve high sulfur concentrations. The development of these materials over the last five to ten years has led to catalysts with controlled activity and selectivity, increased thermal resistance and activity maintenance, and a potential for improved sulfur resistance. The applications which have been briefly mentioned in this last section will be treated in some detail for the specific processes outlined in Part III.

REFERENCES

1. Cecil, R. R., Kmak, W. S., Sinfelt, J. H., and Chambers, C. W., presented at *Adv. Reform. during 37th Mid-year meeting of API Div. Ref.* New York, May, 1972.
2. Schwarzenbek, E. F., "Origin and Refining of Petroleum," p. 94. American Chemical Society, Washington, D.C., 1971.
3. McCoy, C. S., and Munk, P., presented at *Symp. Catal. Reform.* paper 42a, 68th Nat. Meeting of AIChE, Houston, Texas, February 28–March 4, 1971.
4. Sinfelt, J. H., and Barnett, A. E., U. S. Patent No. 3,839,194 (1974).
5. Sinfelt, J. H., Barnett, A. E., and Dembinski, G., U. S. Patent No. 3,442,973 (1968).
6. Kittrell, J. R., U. S. Patent No. 3,576,736 (1971).
7. Inami, S. H., and Wise, H., *J. Catal.* **26**, 92 (1972).
8. Allison, E. G., and Bond, G. C., *Catal. Rev.* **7**, 233 (1972).
9. Cusumano, J. A., U. S. Patent No. 3,844,981 (1974).
10. Flank, W. H., and Beachell, H. C., *J. Catal.* **8**, 316 (1967).

11. Myers, J. W., and Prange, F. A., U. S. Patent No. 2,911,357 (1959).
12. Sinfelt, J. H., Barnett, A. E., and Carter, J. L., U. S. Patent No. 3,617,518 (1971).
13. Sinfelt, J. H., Carter, J. L., and Yates, D. J. C., *J. Catal.* **24**, 283 (1972).
14. Dowden, D. A., *J. Chem. Soc.* 242 (1950).
15. Dowden, D. A., Reynolds, P., *Discuss. Faraday Soc.* **8**, 184 (1950).
16. Schwab, G. M., *Discuss. Faraday Soc.* **8**, 166 (1950).
17. Emmett, P. H., and Skau, N. J., *J. Am. Chem. Soc.* **65**, 1029 (1943).
18. Reynolds, P. W., *J. Chem. Soc.* 265 (1950).
19. Hall, W. K., and Emmett, P. H., *J. Phys. Chem.* **62**, 816 (1958).
20. Van der Plank, P., and Sachtler, W. M. H., *J. Catal.* **12**, 35 (1968).
21. Campbell, J. S., and Emmett, P. H., *J. Catal.* **7**, 252 (1967).
22. Hall, W. K., and Emmett, P. H., *J. Phys. Chem.* **63**, 1102 (1959).
23. Hüfner, S., Wertheim, G. K., Cohen, R. L., and Wernick, J. H., *Phys. Rev. Lett.* **28**, 488 (1972).
24. Seib, D. H., and Spicer, W. E., *Phys. Rev. Lett.* **20**, 1441 (1968).
25. Boudart, M., Cusumano, J. A., and Levy, R. B., New Catalytic Materials For The Liquefaction of Coal, sponsored by the Electric Power Res. Inst. Palo Alto, California, Rep. No. RP-415-1, October 30, 1975.
26. Sinfelt, J. H., *J. Catal.* **29**, 308 (1973).
27. Sinfelt, J. H., and Cusumano, J. A., *in* "Advanced Materials in Catalysis" (J.J. Burton and R. L. Garten, eds.). Academic Press, New York, 1977.
28. Sinfelt, J. H., *Catal. Rev.-Sci. Eng.* **9**, 147 (1974).
29. Sinfelt, J. H., *Chem. Eng. Progr. Symp. Ser.* **63**, 16 (1967).
30. Selwood, P. W., "Adsorption and Collective Paramagnetism." Academic Press, New York, 1962.
31. Bonzel, H. P., and Ku, R., *J. Chem. Phys.* **58**, 4617 (1973).
32. Park, R. L., and Houston, J. E., *J. Vac. Sci. Technol.* **10**, 176 (1973).
33. Sayers, D. E., Lytle, F. W., and Stern, E. A., *Phys. Rev. Lett.* **27**, 1204 (1971).
34. Norris, L. F., and Parravano, G., *in* "Reactivity of Solids" (J. W. Mitchell *et al.*, eds.), p. 149. Wiley (Interscience), New York, 1969.
35. Schlatter, J. C., Sintering of Supported Metals, presented at *Int. Conf. Sinter. and Related Phenomena, 4th* Univ. of Notre Dame, May 26–28, 1975.
36. Slinkin, A. A., and Fedorovskaya, E. A., *Russ. Chem. Rev.* **40**, 860 (1971).
37. Bernstein, L. S., NOx Reduction With Nickel–Copper Alloy and Stabilized Ruthenium Catalysts, preprints, *Jpn.-U.S.A. Sem. Catal. NOx Reactions, Susono, Japan* p. 801, Nov. 2–4, 1975.
38. Burton, J. J., Hyman, E., and Fedak, D., *J. Catal.* **37**, 106 (1975).
39. Ollis, D. F., *J. Catal.* **23**, 131 (1971).
40. Hoffman, D. W., *J. Catal.* **27**, 374 (1972).
41. Sachtler, W. M. H., and Jongepier, R., *J. Catal.* **4**, 665 (1965).
42. Quinto, D. T., Sundaram, V. S., and Robertson, W. D., *Surface Sci.* **28**, 504 (1971).
43. Williams, F. L., and Boudart, M., *J. Catal.* **30**, 438 (1973).
44. Bouwman, R., Lippits, G. J. M., and Sachtler, W. M. H., *J. Catal.* **25**, 350 (1972).
45. Bouwman, R., Lippits, G. J. M., *J. Catal.* **26**, 63 (1972).
46. Bartholomew, C. H., Quart. Tech. Progr. Rep. (April 22–July 22, 1975), ERDA Contract No. E(49-18)-1790, August 6, 1975.
47. Weisz, P. B., *J. Catal.* **6**, 425 (1966).
48. Mills, G. A., Weller, S., and Cornelius, E. B., *Proc. 2nd Int. Congr. Catal. 2nd, Technip. Paris, 1960* p. 2221, 1961.
49. Robertson, S. D., Kloet, S. C., and Sachtler, W. M. H., *J. Catal.* **39**, 234 (1975).

50. Clarke, J. K. A., *Chem. Rev.* **75**, 291 (1975).
51. Hall, W. K., and Hassell, J. A., *J. Phys. Chem.* **67**, 636 (1963).
52. Takeuchi, T., Sakaguchi, M., Miyoshi, I., and Takabatake, T., *Bull. Chem. Soc. Jpn.* **35**, 1390 (1962).
53. Kominami, N., Iwaisako, T., and Ohki, K., German Patents 2,123,606 (1971); 2,127,348 (1971); 2,141,420 (1972).
54. Davies, E. E., Elkins, J. S., and Pitkethly, R. C., German Patent 2,117,651 (1971).
55. Sinfelt, J. H., and Barnett, A. E., U. S. Patent 3,567,625 (1971).
56. Ponec, V., and Sachtler, W. M. H., *J. Catal.* **24**, 250 (1972).
57. Burton, J. J., and Hyman, E., *J. Catal.* **37,** 114 (1975).
58. Sinfelt, J. H., Lam, Y. L., Cusumano, J. A., and Barnett, A. E., *J. Catal.* **42**, 227 (1976).
59. Van Schaik, J. R. H., Dessing, R. P., and Ponec, V., *J. Catal.* **38,** 273 (1975).
60. Ponec, V., and Sachtler, W. M. H., *Proc. Int. Congr. Catal., 5th* p. 645. North-Holland Publ., Amsterdam, 1973.
61. Moyes, R. B., and Wells, P. B., *Adv. Catal.* **23**, 121 (1973).
62. Vannice, M. A., *J. Catal.* **37,** 449 (1975).
63. Vannice, M. A., *J. Catal.* **37**, 462 (1975).
64. Christmann, K., and G. Ertl, *Surface Sci.* **33**, 254 (1972).
65. Thomas, C. L., "Catalytic Processes and Proven Catalysts," p. 104. Academic Press, New York, 1970.
66. Oki, S., Hapel, J., Hnatow, M., and Kaneko, Y., *Catal. Proc. Int. Congr., 5th, 1972* **1**, 173 (1973).

Chapter 3

Effects of Catalyst–Support Interaction, Particle Size, and Surface Morphology

In the last decade the close study of catalyst–support interactions and the role of catalyst morphology and structure in activity has resulted in a better understanding of many catalytic reactions and in the development of new and improved catalysts.

The high surface area of a catalyst required for high activity is obtained by decreasing particle size and is maintained by supporting the small particles on high surface area substrates. The size reduction and the provision of the support can induce physical and chemical changes of the catalyst particles and can alter their catalytic properties. Changes in the particle size of a metal, for example affect the distribution of exposed faces, surface irregularities, steps and edges, and can, particularly in multimetallic systems, lead to phase changes. The chemical changes become more significant as the particle size decreases and are paralleled by increasing interaction of the catalyst with the support and with additives capable of modifying catalytic properties. The same holds true for the support when composed of very small particles (100–200 Å) in order to provide high surface area.

3.1 CATALYST STABILIZATION

One of the most important concerns for catalytic reactions is the decrease of activity due to loss in active surface area or to catalyst volatilization. Exothermic reactions, such as methanation, Fischer–Tropsch, and methanol syntheses, present particularly serious problems, since in these reactions loss of surface area often results in a change of product distribution. Supports can also suffer loss of surface area by sintering under severe reaction or regeneration conditions.

Two approaches for the prevention of loss of surface area will be discussed in the following.

3.1.1 Stabilizing Additives

The use of additives to improve the thermal stability of a catalyst is not new. In Fe ammonia synthesis catalysts, for example, Al_2O_3 has been used as a structural promoter for many years. It is believed to form small domains within the Fe particles, thus preventing agglomeration [1]. Another example of structural promotion is the use of CrO_3 in the Ni-catalyzed dehydrogenation of isopropanol [2]. This type of promotion is probably more prevalent in multicomponent catalysts than is currently recognized.

The possibility of stabilization of alumina supports has been studied by the addition of small amounts of certain oxides which include CuO, MoO_3, V_2O_5, Cr_2O_3, TiO_2, CdO, ZnO, CaO, and MgO [3]. While the addition of CuO, MoO_3, and V_2O_5 leads to a dramatic loss in surface area after heating to 1000°C for 24 h, MgO, SiO_2, and CeO_2 have a stabilizing action and can maintain surface areas as high as 30 $m^2\ gm^{-1}$ at 1200°C. It is believed that stabilization is a consequence of the inhibition of the conversion of the high surface area alumina into α-Al_2O_3. On the other hand, MoO_3 causes surface area loss by forming first aluminum molybdate at 600°C which then decomposes to Al_2O_3 at 800°C. Thus MoO_3 has a catalytic effect on the conversion to alumina which is interestingly inhibited by the presence of Co in the case of alumnia-supported Co–Mo catalysts [4].

In general, it is difficult to predict the direction of the stability change of a material in the presence of other ions. In the TiO_2 system, for example, the anatase–rutile transformation is accelerated considerably by the presence of elements such as Cu, Mn, and Fe [5]. Conversely, the presence of B_2O_3 in Al_2O_3 has been claimed to inhibit the formation of α-Al_2O_3 [6].

In zeolite catalysts the question of stability has been of particular concern, since many zeolite forms are temperature sensitive, especially in the presence of water. The incorporation of rare earths, fluorine, and Mg and the exchange of Na^+ for NH_4^+ in NaY and NaX zeolites result in zeolites stable above 800°C [7, 8]. This is a considerable improvement over the untreated zeolites, whose surface area and structural integrity decreases rapidly above 500°C, particularly in the presence of steam. High stability can also be achieved by careful removal of Al, which leads to a structural change. For example, Kerr [9] found that a hydrogen zeolite, Type Y, heated 2–4 h at 700–800°C in an inert atmosphere where the chemical water remains in the environment of the hydrogen zeolite, yields a new zeolite with remarkable thermal stability, even capable of enduring temperatures as high as 1000°C. He showed that in the new structure approximately 25% of the aluminum was present in the cationic form [10] and found that the removal of aluminum from sodium Y zeolite through the use of dilute solutions of ethylenediaminetetraacetic acid produced a zeolite with improved thermal stability and increased sorptive capacity. Bed geometries which impede the removal of ammonia from ammonium Y zeolite during heating also give an ultrastable zeolite. The mechanism of stabilization is not completely understood, but is attributed to the removal of tetrahedrally coordinated aluminum which causes a contraction of the unit cell and leads to increased structure stability.

3.1.2 Catalyst–Support Interactions

Another important factor in the stabilization of catalysts is the effect of the support on the active catalyst phase. Two approaches have been explored in this regard: the choice of supports or support components to minimize aggregation of the catalytic material, and the use of an additive to eliminate or minimize loss of the catalytic phase during use or regeneration. Both approaches are related inasmuch as they both rely on the chemical interaction of the components.

The stability of a catalyst depends critically on the environment used. Thus, in an oxidizing environment, the stability of a metal such as Ni or Ag on an oxide support is greater than in a reducing environment [11]. This is due to the enhanced interaction of the oxidized surface of the metal with the support. Such a stabilization in an oxidizing environment, however, may be overshadowed by a tendency of the metal oxide to volatilize. Ruthenium catalysts show this effect very strikingly while other noble metals volatilize to a lesser extent. Pt will agglomerate much more severely in an oxidizing environment than in a reducing onc [12, 13]. This is primarily due to the formation of volatile PtO_2. It has also been suggested that in a reducing environment some reduction of the Al_2O_3 surface leads to stronger $Pt–Al_2O_3$ interaction [11]. The degree of this interaction varies with the support. For Ni, for example, Geus showed [11] that the energy of interaction varies from 1.5 kcal g-atom^{-1} for Ni/BeO to 7.3 kcal g-atom^{-1} for Ni/ThO_2. On alumnia it varies with the metal from 3.5 to 6.5 kcal g-atom^{-1} in the series $Ag < Ni < Fe$. As these values are typical physisorption energies, the interactions are rather weak in these cases. Since most reports on the energetics of solid-state interactions are qualitative, it is difficult to compare the above quoted energies with those of other systems.

Russian authors have reported stabilization of supported Pt catalysts by the addition of rare-earth elements [14] or even of Re [15]. In the latter case, which involved both Pt and Pd on SiO_2, the degree of reduction of the Re was not established. Al_2O_3 and SiO_2, have a distinct effect on the dispersion. This has been observed for several metals, including Pt [16] and Ru [17].

Loss of catalyst due to volatilization of the oxide is the primary concern in the case of Ru in an oxidizing environment, in particular in the use for NO_x reduction in automotive exhaust. While the main reaction occurs in a reducing environment, oxidizing transients cannot be avoided. The solution of this problem provides a good example of the application of the solid-state chemistry of the catalyst components and the use of catalyst–support interactions. Two supports have been used: BaO and MgO. They show two different, but related, effects. In the case of BaO, pioneered by Ford workers [18], Ru and BaO interact under oxidizing conditions to form a ruthenate, $BaRuO_3$. Under reducing conditions, Ru returns to the metallic state and is available for the catalytic reaction. Other ruthenate-forming compounds, including the rare-earth oxides, have also been used. The problem with these materials is durability, partly related to cycling between ruthenate and ruthenium metal. The use of MgO overcomes this problem since it does not form a bulk ruthenate. Sufficient surface interaction appears to occur between the finely dispersed Ru and the MgO to inhibit sintering and minimize RuO_4 volatilization [19, 20].

Additives and supports can alter the chemistry of catalytic materials, leading to a change in reduction characteristics of the catalyst and in its activity. In the case of Ni steam-reforming catalysts, for example, reduction is found to occur more readily as the concentration of Ni on the support increases [21, 22]. To achieve reduction, lower H_2/H_2O ratios are required for the unsupported material. A similar effect is observed in the thermal decomposition of chloroplatinic acid. The decomposition of unsupported chloroplatinic acid starts at 100°C, and is completed at 550°C [23], while that of alumina-supported catalysts requires temperatures as high as 760°C [13]. The effect varies on different supports, as seen in the study of the reducibility of chloroiridic acid on various oxides [24]. By following the photoelectron spectra of the iridium–chlorine complex, the reducibility series $ZnO < SiO_2 < TiO_2 < Al_2O_3$ was observed. It is interesting that this is the same series as expected from the acidity of the supports [25].

A change in the catalytic properties of a metallic catalyst as a function of the support has been discussed in the literature for many years, in particular through the work of Schwab and co-workers [26–28]. These authors expected an influence of the electronic population of the oxide support on the metallic catalyst and anticipated a variation in the catalytic properties by varying the electron population of the support by doping. The results of this work showed that, while the effect of these variations was too small to significantly alter the catalytic behavior, there was an effect of dopant on the activation energy for the dehydrogenation of formic acid over nickel, copper, and silver [27]. The effect is readily detectable with thin metal layers. A similar observation was made earlier by Selwood and co-workers [29, 30], who found that compounds deposited on substrates in thin layers are influenced in their structure by the substrate. In several cases the overlayer assumed the substrate structure, and exhibited a valence state that was not the expected stable state of the particular oxide.

The information summarized by Geus [11], and discussed earlier, shows that the effect of an oxide substrate is most significant on an oxide catalyst. This agrees with Selwood's observations. Ross and Delgass compared the properties of europium oxide supported on Al_2O_3 or SiO_2 with those of the unsupported material [31]. The test reaction was the reverse of the water–gas shift reaction, and the chemical state of europium was monitored by Mössbauer spectroscopy. The behavior of the catalyst was dramatically different for the supported and unsupported cases only in one aspect: the effect of CO_2 pressure on the kinetics. Preliminary Mössbauer information suggested the presence of strong support interaction. Similar strong interactions have been reported by Cimino and co-workers [32] in an investigation of N_2O decomposition on a number of solid solutions of the oxides of Co, Cu, and Ni supported on MgO and ZnO. The interpretation of this effect, however, is complicated by the participation of the so-called support in the reaction.

A final example of a support effect on catalytic activity is a methanation study on Raney nickel and Ni supported on ZrO_2 and Al_2O_3 by Dalla Betta and co-workers [33]. The activity per unit surface area of Ni followed the order Ni $>$ $Ni/Al_2O_3 > Ni/ZrO_2$, while the sulfur tolerance showed the reverse order. The significant aspect of this study is that the basis for comparison was the specific activity, i.e., the activity per unit area of nickel. On another basis of comparison, the results would have been meaningless in view of the considerable variation of the dispersion of the catalysts tested. This poses a question with respect to the effect of particle size on the reaction, which topic will be discussed next.

3.2 EFFECTS OF SURFACE STRUCTURE AND PARTICLE SIZE

An important factor that has to be considered in the examination of the catalytic behavior of small particles relates to the change in structure with particle size, an effect that has been recognized only within the last ten years [34, 35]. According to a study of the structure of small particles published by Van Hardeveld and co-workers [36, 37] the surface of a particle departs from its characteristic large-particle morphology when its size is reduced to 100 Å, the most marked changes occurring in the 40–15Å range and affecting particularly the coordination of a surface atom. The equilibrium surface of a large crystallite of an fcc metal, for example, has predominantly 9-coordinated surface sites. An atom adsorbed on such a site has at most 3 nearest neighbors. As the particle size decreases, the number of these sites decreases. A 14-Å crystal, in fact, will have only 33% of such sites [38]. As the number of high-coordination sites decreases, the number of sites which provide 5 nearest neighbors to an adsorbed atom increases to reach a maximum around 20–25 Å [37]. This has a strong effect on the surface properties of the crystallite.

The catalyst particles are not necessarily at equilibrium under reaction conditions. Metastable surface arrangements or surface reconstruction due to impurities or reactants may also lead to modifications in catalytic behavior [39].

Consideration of the effect of surface changes on catalysis led Boudart [38] to divide reactions into two categories: facile and demanding. Facile reactions are those that are not affected by structural changes of the catalyst in the critical region of 20–100 Å. Demanding reactions, on the other hand, require special coordinations or arrangements on the catalytic surface, and therefore depend on the morphology and structure of the catalyst.

The last decade has seen considerable activity in the study of the structure sensitivity of reactions. The interest has centered primarily on the effect of particle size [34, 36, 38, 40, 41–58], but some work was devoted more specifically to surface structure [59–62] and its possible rearrangements with changes in environment or pretreatment [63].

Reactions unaffected by particle size include hydrogen–deuterium exchange and the hydrogenation and dehydrogenation of hydrocarbons [38]. The hydrogenolysis of cyclopentane has also been reported to be a facile reaction [38]; however, most hydrogenolyses are structure sensitive [64, 65]. The reactions of neopentane illustrate the concept of structure sensitivity [56]. In the presence of hydrogen, two parallel reactions are expected: isomerization and hydrogenolysis to smaller hydrocarbon fractions. Several arguments indicate that the isomerization to isopentane requires special sites available only on (111) planes or arrangements of sites leading to such planes [65, 34]. When isomerization occurs on such sites, cracking is minimized. Therefore, changes in particle size that lead to (111) planes are expected to increase the selectivity to isomerization. This effect is indeed observed on supported Pt catalysts [34], and is particularly striking after a heat treatment that leads to a higher density of (111) surfaces. Such surface reconstruction has been reported in the synthesis of ammonia for which the activity of a Fe/MgO catalyst increases after treatment with nitrogen [63]. The occurrence of the effect only in the case of large particles (300 Å) suggests reconstruction of surface sites not present in smaller Fe crystallites.

In the elegant experiments of Somorjai *et al.* [61], the demanding nature of the dehydrocyclization on n-heptane was studied. By using single crystals cleaved systematically to show various well-characterized surfaces and steps, these workers found that, in the presence of hydrogen, the rate of reaction is higher on stepped surfaces of (111) orientation than on corresponding surfaces with (100) orientation. The hydrogenolysis of ethane is also a demanding reaction, as shown by Sinfelt and co-workers for Ni and Rh [66, 67]. For Ni, the specific rate was measured over a number of supported samples with particle sizes increasing from

29 to 88 Å [67]. A decrease in specific activity by a factor of 20 was observed in this range, with the most pronounced decrease occurring between 29 and 57 Å. This is in line with the discussions of Van Hardeveld [37] presented earlier. A similar effect of particle size was observed for Rh [67].

There is preliminary evidence that methanation is also a structure-sensitive reaction [33]. Dalla Betta *et al.* found that the specific steady-state methanation activity of Ru increases with increasing particle size. It is not clear how the variation of poisoning by carbon with increasing particle size affects this conclusion. Studies of the H_2S poisoning of these catalysts, however, tend to support the particle size effect: the introduction of H_2S alters the product distribution by increasing the yield of heavier hydrocarbons. The authors suggest that this is an indication of increased carbon–carbon bond-forming activity with the interruption of the contiguous metallic surface, an effect that would be favored by smaller particles. This phenomenon may have important implications in coal conversion reactions.

3.3 PRACTICAL IMPLICATIONS

The improvement in the thermal stability of supports and the sintering resistance of highly dispersed catalysts has broad implications for most coal conversion related technology, from methanation and Fischer–Tropsch reaction to desulfurization. In fact, catalyst sintering during reaction is one of the most severe methanation problems. On the other hand, support sintering during regeneration, in particular in the presence of steam, is a potential problem of catalysts used in the desulfurization and refining of hydrocarbon liquids.

While the effect of a support on activity and selectivity is often only of secondary importance, there are some reactions in which support change may have significant effects. In methanation, for example, the support has an effect on the sulfur tolerance of Ni-catalysts without altering the specific activity. Such a selective weakening of the interaction of the catalyst with a poison is a very desirable effect, and offers one method for improvement of poison-sensitive catalysts.

According to Sinfelt [64], the effect of particle size on the hydrogenolysis of ethane is a minor contribution compared to the dramatic activity change encountered among different metals. Therefore, the importance of the particle size effect has only indirect practical significance. The difference between demanding and facile reactions, however, is of great significance. This is particularly true in reactions such as hydrogenation, where the undesirable hydrogenolysis reactions are much more structure sensitive. Small variations in the arrangement of sites may, therefore, affect hydrogenolysis considerably with little effect on hydrogenation activity.

The effect of particle size on selectivity is also likely to be important in methanation and Fischer–Tropsch synthesis. The work of Dalla Betta *et al.* [33] suggests a change in higher molecular weight products with decreasing particle size for Ru, a maximum being observed at 24 Å—an effect similar to that reported by Yates and Sinfelt for ethylene hydrogenolysis on rhodium [67]. Thus it appears that, in methanation, particle size affects both selectivity and sulfur tolerance. This clearly is an interesting area for further research. Whether these concepts apply to high-pressure operation remains to be determined, particularly in light of the work of Kreindel *et al.* [68], which suggests that at high pressures the effects of support interaction and particle size on methanation activity are insignificant, and only activity maintenance is important. Sulfur impurities can affect not only the continuity of available sites on a surface, but also surface structure. Thus, the possibility of surface reconstruction, as suggested by Somorjai [39], may be an important factor in the poisoning sensitivity of demanding reactions.

REFERENCES

1. Boudart, M., Topsøe, H., and Dumesic, J. A., *J. Catal.* **28**, 477 (1973).
2. Mears, D. E., and Boudart, M., *AIChE J.* **12**, 313 (1966).
3. Gaugin, R., Graulier, M., and Papee, D., *in* Catalysts for the Control of Automotive Pollutants (J. McEnvoy, ed.), p. 147. American Chemical Society, Washington, D.C., 1975.
4. Ratnasamy, P., Mehrotra, R. P., and Ramaswamy, A. V., *J. Catal.* **32**, 63 (1974).
5. MacKenzie, K. J. D., *Trans. J. Brit. Ceram. Soc.* **74**, 77 (1975).
6. McArthur, D. P., U. S. Patent No. 3,883,442 (1975).
7. Elliott, C. H., Jr., German Offen. 2,227,334, June 7, 1971; U. S. Appl. 150,780, June 7, 1971.
8. Kerr, G. T., *in* "Molecular Sieves" (W. M. Meier and J. B. Uytterhveven, eds), p. 219. American Chemical Society, Washington, D.C., 1973.
9. Kerr, G. T., *J. Phys. Chem.* **73**, 2780 (1969).
10. Kerr, G. T., *J. Phys. Chem.* **71**, 4155 (1967).
11. Geus, J. W., *Int. Symp. Sci. Basis Catalyst Preparation, Brussels.* October 14–17, 1975.
12. Baker, R. T. K., Thomas, C., and Thomas, R. B., *J. Catal.* **38**, 510 (1975).
13. Mills, G. A., Weller, B., and Cornelius, E. B., *Proc. Int. Congr. Catal. 2nd, Technip, Paris, 1960.* p. 2221, 1961.
14. Kozlov, N. S., Sen'kov, G. M., Zaretskii, M. V., Davidovskaya, A. M., and Palei, S. V., *Dokl. Akad. Nauk SSSR* **18**, 621 (1974).
15. Bursian, N. R., Kogan, S. B., Semenov, G. I., and Levitskii, E. A., *Kinet. Katal.* **15**, 1608 (1974).
16. Cusumano, J. A., Dembinski, G. W., and Sinfelt, J. H., *J. Catal.* **5**, 471 (1966).
17. Vedenyapin, A. A., Klabunovskii, E. I., Talanov, Yu, A., and Sokolova, N. P., *Kinet. Katal.* **16**, 436 (1975).
18. Shelef, M., and Gandhi, H. S., *Platinum Metals Rev.* **18**, 2 (1974).
19. Deluca, J. P., Murell, L. L., Rhodes, R. P., and Tauster, S. J., presented at American Chemical Society Annual Meeting, Chicago, Illinois, 1975.
20. Bernstein, L. S., *Japan-U.S.A. Seminar Catal. NOx React., Susono, Japan* Nov. 2–4, 1975.

21. Rostrup-Nielsen, J. R., "Steam Reforming Catalysts," p. 42. Danish Tech. Press, Copenhagen, 1975.
22. Ross, J. R. H., and Steel, M. C. F., *Trans. Faraday Soc.* **69**, 10 (1973).
23. Shubochkin, L. K., Gushchin, V. I., Larin, G. M., and Kolosov, V. A., *Zh. Neorg. Khim.* **19**, 460 (1974).
24. Escard, J., Pontvianne, B., and Contour, J. P., *J. Electron Spectrosc. Relat. Phenom.* **6**, 17 (1975).
25. Tanabe, K., "Solid Acids and Bases." Academic Press, New York, 1970.
26. Schwab, G. M., *Angew. Chem.* **73**, 399 (1961).
27. Schwab, G. M., Block, J., and Schultze, G., *Naturwissenschaften* **44**, 482 (1957); *Angew. Chem.* **71**, 101 (1959).
28. Schwab, G. M., and Mutzbauer, G., *Z. Phys. Chem.* **32**, 367 (1962).
29. Hill, H. F., and Selwood, P. W., *J. Am. Chem. Soc.* **71**, 2522, (1949).
30. Selwood, P. W., *Bull. Soc. Chim. Fr.* No. 3-4, 167 (1949).
31. Ross, P. N., Jr., and Delgass, W. N., *J. Catal.* **33**, 219 (1974).
32. Cimino, A., Pepe, F., and Schiavello, M., *in* "Catalysis" (J. W. Hightower, ed.), p. 125. North-Holland Publ., Amsterdam, 1973.
33. Dalla Betta, R. A., Piken, A. G., and Shelef, M., *J. Catal.* **40**, 173 (1975).
34. Boudart, M., Aldag, A., Ptak, L. D., and Benson, J. E., *J. Catal.* **11**, 35 (1968).
35. Poltorak, O. M., and Boronin, V. S., *Zh. Fis. Khim.* **40**, 2671 (1966).
36. Van Hardeveld, R., and Hartog, F., *Catal.* **22**, 75 (1972).
37. Van Hardeveld, R., and Hartog, F., *Surface Sci.* **15**, 189 (1969).
38. Boudart, M., *Adv. Catal.* **20**, 453 (1969).
39. Somorjai, G. A., *J. Catal.* **27**, 453 (1972).
40. Cece, J. M., and Gonzalez, R. D., *J. Catal.* **28**, 260 (1973).
41. Primet, M., Basset, J. M., Garbowski, E., and Mathieu, M. V., *J. Am. Chem. Soc.* **97**, 3655 (1975).
42. Pusateri, R. J., Katzer, J. R., and Manogue, W. H., *AIChE J.* **20**, 219 (1974).
43. Ostermaier, J. J., Katzer, J. H., and Manogue, W. H., *J. Catal.* **33,** 457 (1974).
44. Corolleur, C., Gault, F. G., Juttard, D., Maire, G., and Muller, J. M., *J. Catal.* **27**, 466 (1972).
45. Dautzenberg, F. M., and Plateeuw, J. C., *J. Catal.* **24**, 364 (1972).
46. Dumesic, J. A., Topsøe Khammouma, S., and Boudart, M., *J. Catal.* **37,** 503 (1975).
47. Luss, D., *J. Catal.* **23**, 119 (1971).
48. Cha, D. Y., and Parravano, G., *J. Catal.* **18**, 200 (1970).
49. Carrà, S., and Ragaini, V., Structure and Catalytic Behavior of Supported Metals, presented at the *Int. Congr. "Chemistry Days" Ind. Catal., 20th, Milan* May 19–21, 1969.
50. Bond, G. C., *Proc. Int. Congr. Catal., 4th, Moscow, 1968* Paper 67, p. 266 (1971).
51. Anderson, J. R., and Shinioyama, Y., *in* "Catalysis" (J. W. Hightower, ed.), Vol. 1, p. 695. North-Holland Publ., Amsterdam and American Elsevier, New York, 1973.
52. Coenen, J. W. E., Van Meerten, R. Z. C., and Rijnten, H. th., *in* "Catalysis" (J. W. Hightower, ed.), Vol. 1, p. 671. North-Holland Publ., Amsterdam and American Elsevier, New York, 1973.
53. Oliver, R. G., and Wells, P. B., *in* "Catalysis" (J. W. Hightower, ed.), Vol. 1, p. 659. North-Holland Publ., Amsterdam and American Elsevier, New York, 1973.
54. Maire, G., Corolleur, O., Juttard, D., and Gault, F. G., *J. Catal.* **21**, 250 (1971).
55. Gustafson, W. R., and Kapner, R. S., presented at the *Symp. Recent Adv. Kinetc. Catal.* Part II, AIChE 60th Annual Meeting, New York, 1967.
56. Boudart, M., and Peak, L. D., *J. Catal.* **16**, 90 (1970).
57. Presland, A. E. B., Price, G. L., and Trimm, D. L., *J. Catal.* **26**, 313 (1972).

58. Clarke, J. K. A., McMahon, E., and O Cinneide, A. D., *in* "Catalysis" (J. W. Hightower, ed.), Vol. 1, p. 685. North-Holland Publ., Amsterdam and American Elsevier, New York, 1973.
59. Kahn, D. R., Peterson, E. E., and Somorjai, G. A., *J. Catal.* **34**, 294 (1974).
60. Boudart, M., Delbouille, Derouane, V., and Walters, A. B., *J. Am. Chem. Soc.* **94**, 6622 (1972).
61. Somorjai, G. A., Joyner, R. W., and Lang, B., *Proc. Roy. Soc.* **331**, 335 (1972).
62. Derouane, E. G., Indovina, V., Walters, A. B., and Boudart, M., *in* "Reactivity of Solids," p. 703. Chapman and Hall, London, 1972.
63. Dumesic, J. A., Topsøe, H., and Boudart, M., *J. Catal.* **37**, 313 (1975).
64. Sinfelt, J. H., *Catal. Rev.* **3**, 175 (1969).
65. Anderson, J. R., and Avery, N. R., *J. Catal.* **5**, 449 (1966).
66. Carter, J. L., Cusumano, J. A., and Sinfelt, J. H., *J. Phys. Chem.* **70**, 2257 (1966).
67. Yates, D. J. C., and Sinfelt, J. H., *J. Catal.* **8**, 348 (1967).
68. Kreindel, A. I., Sobolevskii, V. S., Golosman, E. Z., and Yakerson, V. I., *Kinet. Katal.* **15**, 408 (1974).

Chapter 4

Characterization

4.1 INTRODUCTION

Improved catalyst characterization has contributed to the better understanding of catalytic phenomena and to the development of new concepts in catalysis in the last decade and played a key part in catalyst quality control, activity maintenance, and regeneration. The problem of characterization is complex for several reasons. Most catalysts contain several components that have individual functions and therefore require individual identification. The catalyst particles are often so small that they are beyond the reach of standard materials-analysis equipment. Furthermore, heterogeneous catalysts involve surface phenomena for the characterization of which bulk parameters have only limited utility.

The physical and chemical properties which can be measured in spite of these difficulties include the surface area of individual components of a catalyst (specific surface area), surface acidity, surface composition, and, in certain cases, microscopic surface morphology. The determination of total surface area, pore volume, and pore size distribution is now practiced routinely [1], and requires no discussion.

Surface composition and morphology of materials of catalytic interest have been extensively investigated over the last decade by techniques originally designed to study the electronic properties of the surface and bulk of solids. These methods will be treated in Part II, while the present section will cover specific surface area and surface acidity.

4.2 SURFACE AREA AND PARTICLE SIZE

Determination of the specific surface area of a catalyst involves methods that can distinguish between the surface areas of the support and of the active catalyst materials and include x-ray diffraction [2], transmission electron microscopy [2, p. 363], and gas chemisorption.

Using x-ray diffraction the width of a line for a particular structure can be related to the crystallite size of the compound which can be calculated by a routine procedure. The technique has two limitations, both of which become apparent at low catalyst loadings. One is sensitivity, which is lowered by the support material interfering with the x-ray lines of interest, especially at low concentrations. The other limitation is that particles smaller than 30–40 Å cannot be detected—a serious constraint in view of the large number of catalysts of interest having a considerable fraction of particles smaller than this limit.

The procedure for particle size determination by electron microscopy is also straightforward and routine. Complications are primarily related to sample preparation (to allow transmission) and resolution. The latter is in the 10–20-Å range, although even this resolution requires skillful operation and a sensitive instrument. Some problems are also encountered in obtaining truly representative data.

Because of the limitation of these techniques, the selective adsorption of gases has been the most common and effective means for the measurement of specific surface area. The technique relies on the difference between chemical and physical adsorption. Gases are used which selectively chemisorb on only one component of the catalyst.

4.3 SURFACE AREA BY CHEMISORPTION

The use of gas adsorption for surface area determination of catalysts has been investigated primarily for metals. Other catalytic materials have received only limited attention.

4.3.1 Metals

As can be seen from Table 4-1, most of the metals of catalytic interest have been characterized by the selective adsorption of a number of gases. The subject has been reviewed recently [18], and therefore will be discussed briefly in this section.

TABLE 4-1
SURFACE AREA MEASUREMENT OF METAL CATALYSTS

Metal	Chemisorbed gas	Conditions °C	Conditions Torr	Reference
Fe	H_2	25	100	3
	CO	25	100	3
Co	H_2	25	100	4
	CO	−195		5
Ni	H_2	25		4, 6, 7
	CO	25		6
Cu	H_2	25		4, 8
	N_2O	90	200	9
		25	500	10
	CO	25	100	4
Ru	H_2	25		11
	CO	25		12
Rh	H_2	25		13
	O_2	−80 to 300		13
	CO	25		13
	H_2/O_2	25		13
Pd	H_2	70	1	14
	H_2/O_2	100	< 350	15
	D_2/O_2	100	< 1000	15
Ag	O_2	180	10	16
	N_2O	150		16
Os	H_2	25		17
	CO	25		3
Ir	H_2	25		11
	CO	25		11
Pt	H_2	25		13
	CO	25		13
	H_2/O_2	25		13
	O_2	25		13

The most common sorbate is hydrogen. It is used extensively to determine the surface areas of Pt [19, 20], Ni [4, 6, 7], Co [4], Fe [3], Os, Ir, Rh, and Ru [17], Cu [4, 8], and, under special conditions, of Pd [14]. Its only drawback is the need for careful reduction of the surface and subsequent desorption of the reducing gas.

To overcome this problem, Benson and Boudart [21] developed a technique whereby hydrogen is used to "titrate" oxygen atoms preadsorbed on the metal surface. As indicated by Dalla Betta [22], this procedure eliminates the complications of chemisorption, and increases the sensitivity of the technique. The technique can be used for those metals that catalyze the H_2/O_2 reaction, which include Pt, Pd, Rh, and Ir. The titration has been studied most extensively on Pt [21–25] and has also been applied recently to Pd [15] and Rh [13]. Interpretation of the measurement on Rh is complicated by uncertainties in the stoichiometry of the hydrogen–rhodium surface species and by partial bulk oxidation [13]. The uncertainty in stoichiometry, while not as great for the other metals, is an important factor in the titration of Pt at high dispersions (small particle size) for which changes in the oxygen–hydrogen ratio have been reported [24, 25]. Chemisorption of H_2 is therefore still the most reliable procedure for highly dispersed catalysts. In the case of palladium, formation of a bulk hydride is avoided by operation at higher temperatures (100°C) and the use of D_2 to avoid formation of the β-Pd-H phase [15].

Carbon monoxide is used in preference to hydrogen in the case of Cu [5, 8] on which H_2 adsorption is slow. However, in all cases the primary limitation of the use of CO is its ability to adsorb either in a linear or in bridge mode, leading to CO–metal ratios greater than 1:1 on small particles and to uncertainties about the stoichiometry [4, 12]. Furthermore, considerable CO adsorption occurs on the support, thus requiring corrections that are often as high as the chemisorption on the metal in dilute systems.

Oxygen chemisorption has had only limited application, since many of the metals of interest form bulk oxides at or above room temperature. It has been used extensively for the determination of the surface area of silver catalysts at 175–200°C [16] to minimize formation of silver oxide which is unstable at this temperature.

An alternate technique of oxygen chemisorption involves the use of N_2O as an oxygen source, via the decomposition to N_2 and adsorbed oxygen. It has been used for the measurement of the surface area of Cu [9, 10], and of Ag [16]. Surprisingly N_2O does not cause bulk oxidation below 100°C in the case of copper [9], apparently because of kinetic effects.

4.3.2 Oxides

In sharp contrast to metals, the characterization of oxides by selective gas chemisorption is much more limited, making the measurement procedures complex. The most common adsorbent used is CO. The dependence of the adsorption isotherm on sample preparation and pretreatment (reported, for example, for Co_3O_4/SiO_2 [26]) makes CO adsorption data often unreliable.

Shelef and co-workers used NO chemisorption to measure the surface area of oxides of iron [27], chromium [28], copper [29], and nickel [30]. The technique relies on the observation that the adsorption of NO on a number of oxides follows a Freundlich-type isotherm. This behavior permits calculation of monolayer coverage by plotting the logarithm of the gas uptake against the logarithm of the equilibrium pressure at various temperatures. It is a promising technique for oxide characterization, and should be extended to other systems.

An oxide catalyst that is of great importance for coal conversion is cobalt molybdate supported on Al_2O_3. Most gas adsorption studies on this catalyst have been directed at the chemistry and kinetics of the interaction [31–34], but there have been a few attempts to measure the specific surface area [35–37]. In one of these the thermal desorption of hydrogen from $CoO–MoO_3–Al_2O_3$ catalysts was compared with thiophene hydrogenolysis activity. Of the various hydrogen bonding states observed, only the weakly bonded hydrogen showed a relationship to thiophene hydrogenolysis. It is unlikely, therefore, that hydrogen adsorption will become a routine technique for the characterization of this catalyst. Other gases, including H_2S and NO, should be explored.

4.4 SURFACE ACIDITY

The acidity of oxides has been studied extensively over the last two decades. Parameters that have been measured include acid strength, acid amount, acid type (Brønsted or Lewis), and acid distribution. Measurement techniques are routine and reproducible. They have been described in detail in a number of recent review articles [38–41], and will therefore not be discussed in this section. Since surface acidity is an important aspect of the characterization of catalysts for coal conversion and particularly for liquefaction and upgrading of coal liquids [42], it will be covered in Part III.

4.5 IMPACT OF CHARACTERIZATION IN CATALYSIS

Comparisons of catalytic activities of materials are meaningless without specific surface area measurements. This was shown quite clearly by Vannice [43] in the re-examination of the methanation activity of transition metals. Prior activity data reported the following activity sequence [44]: Ru > Ir > Rh > Ni > Co > Os > Pt > Fe > Pd. By contrast, on relating the data to unit metal surface Vannice found the series Ru > Fe > Ni > Co > Rh > Pd > Pt > Ir. The most striking difference was found for Fe, which was considered by the early workers to be a poor methanation catalyst. Thus the real difficulty consists in the preparation and stabilization of high surface area Fe catalysts [45], and there is a need for finding appropriate methods.

Another example of systematic comparisons of activity that rely on careful specific surface area measurements is the extensive work of Sinfelt and collaborators on the hydrogenolysis and hydrogenation activity of transition metals [17, 46]. It revealed striking differences in the activity of metals for these two reactions and contributed materially to the understanding of their behavior in applied catalysis.

Characterization has also been responsible for the development of such new ideas and concepts as catalyst–support interactions, structural and chemical promotion, catalyst stabilization, and structure sensitivity. The evaluation of new catalytic materials may be misleading in the absence of selective surface area information. For a long time it was believed that nickel boride was a more active hydrogenation catalyst than nickel and that chromia increased this activity considerably [47]. Mears and Boudart showed that, on a Ni surface area basis, Ni_2B and Ni have the same activity [48]. The chromia "promoter" had only a surface effect: the activity per unit surface area of unpromoted and promoted nickel boride catalysts was constant over a sevenfold range of surface areas.

The elucidation of such surface area stabilization, poisoning, and sintering effects is an important aspect of coal conversion catalysis. It is therefore surprising that such limited attention has been given to the characterization of oxide catalysts, in particular in terms of selective surface area. In the case of HDS and HDN catalysts deactivation studies have been limited to studies of the changes in total surface area and pore volume distribution. Effects of selective poisoning, sintering, and possible chemical change of the active catalytic surface have not been investigated. Lack of catalyst characterization has limited the understanding of the effect of promoters on the activity.

REFERENCES

1. Gregg, S. J., and Sing, K. S. W., "Adsorption, Surface Area and Porosity." Academic Press, New York, 1967.
2. Anderson, J. R., "Structure of Metallic Catalysts," p. 364. Academic Press, New York, 1975.
3. Sinfelt, J. H., and Yates, D. J. C., *J. Catal.* **10**, 362 (1968).
4. Sinfelt, J. H., Taylor, W. F., and Yates, D. J. C., *J. Phys. Chem.* **69**, 95 (1965).
5. Anderson, R. B., Hall, K. W., and Hofer, C., *J. Am. Chem. Soc.* **70**, 2465 (1948).
6. Brooks, C. S., and Christopher, G. L. M., *J. Catal.* **10**, 211 (1968).
7. Yates, D. J. C., Taylor, W. F., and Sinfelt, J. H., *J. Am. Chem. Soc.* **86**, 2996 (1964).
8. Suzuki, M., and Smith, J., *J. Catal.* **21**, 336 (1971).
9. Scholten, J. J. F., and Konvalinka, J. A., *Trans. Faraday Soc.* **65**, 2465 (1969).
10. Osinga, Th. J., Linsen, B. G., and Van Beek, W. P., *J. Catal.* **7**, 277 (1967).
11. Sinfelt, J. H., and Yates, D. J. C., *J. Catal.* **8,** 82 (1967).
12. Dorling, T. A., and Moss, R. L., *J. Catal.* **7**, 378 (1967).
13. Wanke, S. E., and Dougharty, N. A., *J. Catal.* **24**, 367 (1972).
14. Aben, P. C., *J. Catal.* **10**, 224 (1968).
15. Benson, J. E., Hwang, H. S., and Boudart, M., *J. Catal.* **30,** 146 (1973).
16. Scholten, J., Knovolinka, J., and Beekman, F., *J. Catal.* **28,** 209 (1973).
17. Dalla Betta, R. A., Cusumano, J. A., and Sinfelt, J. H., *J. Catal.* **19,** 343 (1970).
18. Farrauto, R. J., Determination and Applications of Catalytic Surface Area Measurements, presented at 77th National AIChE Meeting, Pittsburg, Pennsylvania, June 2–5, 1974.

19. Vannice, M. A., Benson, J. E., and Boudart, M., *J. Catal.* **16**, 348 (1970).
20. Hunt, C. E., *J. Catal.* **23**, 93 (1971).
21. Benson, J. E., and Boudart, M., *J. Catal.* **4**, 704 (1965).
22. Dalla Betta, R. A., *J. Catal.* **31**, 143 (1973).
23. Wilson, G., and Hall, K. W., *J. Catal.* **17**, 190 (1970).
24. Wilson, G., and Hall, K. W., *J. Catal.* **24**, 306 (1972).
25. Mears, D. E., and Hansford, R. C., *J. Catal.* **9**, 125 (1967).
26. Pope, D., Walker, D. S., Whalley, L., and Moss, R. L., *J. Catal.* **31**, 335 (1973).
27. Otto, K., and Shelef, M., *J. Catal.* **18**, 184 (1970).
28. Otto, K., and Shelef, M., *J. Catal.* **14**, 226 (1969).
29. Gandhi, H., and Shelef, M., *J. Catal.* **28**, 1 (1973).
30. Gandhi, H., and Shelef, M., *J. Catal.* **24**, 241 (1972).
31. Samuel, P., and Yeddanapalli, L. M., *J. Appl. Chem. Biotechnol.* **24**, 777 (1974).
32. Aptekar, E. L., Bykhovskii, M. Ya, Krylov, O. V., Lebedev, Yu A., and Miroshnichewco, E. A., *Kinet. Katal.* **15**, 1568 (1973).
33. Kawak, M., and Krajewski, J., *Zesz. Nauk. Politech. Slask. Chem.* **61**, 11 (1973).
34. Slager, T. L., and Amberg, C. H., *Can. J. Chem.* **50**, 3416 (1972).
35. Kolosov, A. K., Shvets, V. A., and Kazanskii, V. P., *Kinet. Katal.* **16**, 197–201 (1975).
36. Dollimore, D., Galwey, A., and Rickett, G., *J. Clim. Phys. Phys.-Chem. Biol.* **72,** 1059 (1975).
37. Masoth, F. E., *J. Catal.* **36**, 164 (1975).
38. Goldstein, M. S., *in* "Experimental Methods in Catalytic Research" (R. B. Anderson, ed.), Chapter 9. Academic Press, New York, 1968.
39. Tanabe, K., "Solid Acids and Bases." Academic Press, New York, 1970.
40. Donnet, J. B., *Bull. Soc. Chim. Fr.* 3353 (1970).
41. Forni, L., *Catal. Rev.* **8,** 65 (1973).
42. Boudart, M., Cusumano, J. A., and Levy, R. B., New Catalytic Materials for the Liquefaction of Coal, Rep. RP-415-1, Electric Power Research Institute, October 30, 1975.
43. Vannice, M. A., *J. Catal.* **37**, 449 (1975).
44. Fischer, F., Tropsch, H., and Dilthey, P., *Brennst.-Chem.* **6**, 265 (1925).
45. Boudart, M., Delbouville, A., Khammouma, S., and Topsøe, H., *J. Catal.* **37**, 486 (1975).
46. Sinfelt, J. H., *Catal. Rev.* **3**, 175 (1969).
47. Paul, R., Buisson, P., and Joseph, N., *Ind. Eng. Chem.* **44**, 1006 (1952).
48. Mears, D. E., and Boudart, M., *AIChE J.* **12**, 313 (1966).

Chapter 5

Catalyst Preparation

5.1 INTRODUCTION

For many years catalyst preparation has been regarded as an art, and most of the available information was contained in the patent literature and a few review articles [1–5]. In recent years, there has been an increased effort to put this activity on a scientific basis as shown by the first international conference devoted to this subject [6]. Many new techniques have been developed for catalyst characterization, and these have played an important role in understanding the effects of preparative variables on catalytic properties.

This increased insight into the nature of catalyst preparation has led to the synthesis of a number of new catalysts with improved activity, selectivity, and stability properties. Many of these improvements may well find application in the catalytic conversion of coal. A review of all the details of the numerous new concepts in catalyst preparation would be beyond our scope. Instead, most of the important techniques discovered or developed over the last decade will be mentioned with adequate references. Those techniques which have obvious or potential applications, particularly in the conversion of coal to synthetic fuels, will be discussed in some detail.

A summary of the primary concepts of interest with some examples, as well as the significance of these concepts and areas for possible application, are pre-

TABLE 5-1
CONCEPTS IN CATALYST PREPARATION

Preparative objective	Technique/material	Examples	References	Significance	Areas of applications
Dispersed materials	Chem. reduction	$PtAu_3$ (100 Å)	7–11	Preparation of high surface area unsupported metals, alloys, oxides, mixed oxides, and complex inorganic materials allows catalytic evaluation of well characterized novel materials	Fundamental catalytic studies of novel materials; Broad applications
	Flame decomp.	ZrO_2, Fe_2O_3 (100 Å)	12		
	Flame-spraying	Raney Ni (100 m^2 gm^{-1})	19		
	Support solub.	Pt (15 Å)	13		
	Impreg. synth.	$Mg_2Mo_3O_8$ (100 m^2 gm^{-1})	74		
	Ketenide decomp.	Ag clusters	78		
	Freeze-drying	W_2C (200 Å)	14–18		
Supported catalysts	Incip. impreg.	Supp. metals	53, 54	Variable dispersion of active phase gives control of catalytic properties, anchored complexes obviate separation problems	CO/H_2 synthesis; Water–gas shift; Refining and upgrading
	Immis. liq. displ.	Pt/Al_2O_3	59		
	Surf. impreg.	Ni/Al_2O_3	58		
	Adsorption	$Pt(NH_3)_4Cl_2/Al_2O_3$	60, 61		
	Anchoring	Rh(acac)(CO)-SiO_2	30		
Controlled physical property supports	Sol–gels	$TiO_2 \cdot SiO_2$	23–25	Supports with controlled areas, acidity, pore sizes, and mechanical properties can be made; homogeneous pH control gives high dispersion of supported metals which are difficult to disperse (e.g., Ni, Co, Fe)	Direct liquefaction; CO/H_2 synthesis; Water–gas shift
	Gel precip.	$ZrO_2 \cdot TiO_2$	23–26		
	Homo. pH change	Ni/SiO_2 (13 Å)	62, 63		
	Aerogels	Ni/Al_2O_3	27–29		
	Pyrogels	$BaTiO_3$	23		

TABLE 5-1 *(continued)*

Preparative objective	Technique/material	Examples	References	Significance	Areas of applications
Stabilized supports	Doping	Al_2O_3 w/Cs_2O, La_2O_5	31, 32	Supports are available which can endure 1000–1500°C; ultrastable zeolites and Al_2O_3 are readily prepared; monoliths open new dimension for catalysts in terms of heat/mass transfer, mechanical properties, pressure drop, and other engineering variables	Direct liquefaction
	Stab. zeolites	Decat. Y-Type	33–44		CO/H_2 synthesis (methanation)
	Monoliths	Cordierite	49–52		Refining and upgrading
	Refractories	Si_3N_4, SiC, ZrO_2	45–48		
Stabilized supported metals	Metal/supp. eff.	Ru/MgO, Ru/BaO	69, 70	Metal catalysts which can endure up to 1000°C can be prepared; they exhibit good activity maintenance after oxidative regeneration	CO/H_2 synthesis (methanation)
	Bimetallics	PtCo	55, 56, 68		Refining and upgrading
	Trimetallics	Ru-Ni-Cu	71		
	Sulfidation	H_2S-Pt/monolith	73		
Multimetallic catalysts	Clusters	$RuCu/SiO_2$	55, 56	Control of catalyst activity, selectivity, maintenance, stability, and *possibly* poisoning	CO/H_2 synthesis
	Supp. alloys	$PdAu/SiO_2$	57, 68		Refining and upgrading
	Raney alloys	CoNi, NiCu	20–22		
	Organomet. clusts.	$RhCo_2/SiO_2$	65, 66		

sented in Table 5-1. The table gives a listing of over thirty preparative techniques grouped into six categories.

Many can be used commercially. Others, particularly those concerned with the preparation of dispersed materials, may only be practical on a laboratory scale, but are useful in providing new high surface area catalysts for expeditious evaluation.

5.2 METALLIC CATALYSTS

Metallic catalysts have played a major role in the development of petroleum and petrochemical processes and are also expected to be important for certain aspects of the catalytic conversion of coal.

5.2.1 Unsupported Metal Catalysts

Among the many procedures for preparing unsupported metals only a few provide practical catalysts for the coal conversion reactions of interest. The details of these procedures have been discussed in a recent review [5]. One common technique is chemical reduction of aqueous [7] or nonaqueous [8] solutions of metal salts. Chemical reducing agents include sodium borohydride [7, 8], hydrazine [9], formaldehyde [10], or hypophosphorous acid [10]. High surface area alloys can also be obtained by this procedure [11].

Highly dispersed metals and alloys may be prepared by flame decomposition [12]; i.e., by directing a vapor of a metal salt, such as a metal chloride, into the flame of an oxygen–hydrogen burner, followed by hydrogen reduction of the oxide produced to the metal. Alternate procedures include dissolution of the support in a supported metal catalyst in HF or HBF_4 [13], and freeze-drying of an aqueous metal salt solution with subsequent hydrogen or chemical reduction of the finely dispersed salts [14–18].

Flame-spraying [19] has been used extensively by the Bureau of Mine workers [19] to coat heat-exchanger tubes with nickel catalysts for use as a methanation reactor. Direct contact of the catalyst with the exchanger tubes assures high heat transfer rates, which is reportedly helpful in minimizing catalyst deactivation during the very exothermic methanation reaction. Another method for producing unsupported catalysts is the preparation of Raney metal alloys [20, 21], a recent updating of the well-known Raney Ni technique [22]. These alloys present an opportunity to use the concepts of bimetallic catalysis to improve catalytic properties, poison tolerance, and resistance to thermal degradation.

5.2.2 Supported Metal Catalysts

As metal and support in a supported catalysts generally fulfill markedly different functions, it is convenient to treat them separately.

The preparation of supports with controlled surface area, pore volume, and pore-size distribution has been the subject of much research over the past years. The primary objective was the development of stable, controlled-pore supports for use in hydrofining residua for which conventional HDS/HDN catalysts have proved to be ineffective because of easy plugging of their small pores.

Research in preparing novel supports has included a number of precipitation and gelation procedures which are broadly applicable to oxide or mixed oxide systems. In the sol–gel process [23] a colloidal sol of metallic oxides or hydroxides is converted to a semirigid gel by removal of water, neutralization with base, or solvent extraction of the acid component. The gel is then dried and calcined to the oxide. The final surface area, pore-size distribution, and structure are determined during the gelation stage. In recent years, considerable effort has been devoted to exploring the utility of sol–gel technology for preparing metal oxides [24, 25] for use in fuel rods or other nuclear materials. As a result of this work, it is now possible to prepare such materials as Al_2O_3, TiO_2, ZrO_2, Cr_2O_3, Fe_2O_3, rare-earths, and their mixtures, with well-controlled physical properties.

In gel precipitation an organic gelling agent is added to the aqueous metal salt solution before precipitation. On contact with an alkaline precipitant (NH_4OH), a coprecipitate is formed in which the metallic hydroxide is held rigidly within the framework of the organic gel. By controlling process parameters, one can alter the physical properties of the oxide or mixed oxide support material [23, 73].

According to an interesting technique described by Teichner and co-workers [26–28], mixed nickel-oxide–alumina-aerogel is prepared by adding a small amount of water to a methanol–butanol solution containing dissolved aluminum and nickel compounds (e.g., aluminum *sec*-butylate and nickel acetate). The solvent is then stripped in an autoclave under supercritical conditions, to avoid collapse of the solid texture. On flushing the autoclave with H_2 at high temperatures, metallic nickel on an alumina aerogel is obtained. If inert gas is used, a mixed oxide aerogel is the product. By adjusting parameters, one can obtain substantial variations in surface area and pore volume. Thus nickel oxide can be made with a surface area range of 500–800 $m^2\ gm^{-1}$, also $NiO \cdot Al_2O_3$ (160–650 $m^2\ gm^{-1}$), MoO_2 (90–287 $m^2\ gm^{-1}$), $NiO \cdot MoO_2$ (210–483 $m^2\ gm^{-1}$), and many other combinations. The general procedure has two primary advantages: it produces materials which need little, if any, purification (uses no Cl^-, NO_3^-, or Na^+) and it is extremely quick (~2 h to prepare a few hundred grams in a typical autoclave). The mixed oxides prepared this way have also been found to have interesting catalytic properties [26].

The use of gel procedures permits the preparation of mixed oxides with controlled surface acidity by controlling the composition of the bulk, and therefore that of the surface. As surface acidity can determine product distribution, activity, and its maintenance [29], gel procedures are of interest to synthetic fuels technology.

Techniques for preparing support materials with increased thermal stability are particularly important for catalysts subject to hot spotting (e.g., methanation) or requiring oxidative regeneration (e.g., direct liquefaction).

One procedure for imparting thermal stability to support materials is cationic doping. Thus alumina needs stabilization to prevent its high temperature conversion to the α-form by which its surface area is usually reduced from about 250 m^2 gm^{-1} to less than 1 m^2 gm^{-1}. If alumina is doped with small amounts of oxides from Group IIA (e.g., CaO, SrO, BaO) [30] or of the rare-earths (e.g., CeO_2, La_2O_3) [31] and subsequently calcined at 1200°C for 2 h, a stable surface area of 20–100 m^2 gm^{-1} is obtained for use as a thermally stabilized support material. These materials have found application in automotive exhaust catalysts and catalytic combustion.

The development of zeolites as catalysts and supports has been an important chapter in heterogeneous catalysis, because of the broad range of properties of these materials and their diversified applications to numerous petroleum and petrochemical processes. They can be synthesized in a variety of compositions and readily undergo ion exchange with catalytic metals (e.g., Pt, Pd, Ni). Most of these properties and their implications have been reviewed elsewhere [29]. A major accomplishment was the development of a new class of zeolites which have remarkable thermal and chemical stability [32–38] and can withstand hydrothermal conditions at high temperatures, in some cases exceeding 1000°C, with no structural collapse. Their thermal stability and broad range of physical and catalytic properties make these materials attractive for a number of coal conversion reactions.

Two particularly interesting applications for these stabilized zeolites are in hydrodesulfurization and hydrodenitrogenation catalysts [39–43]. Such catalysts are prepared by a coprecipitation of the molecular-sieve-containing hydrogel composite in an aqueous solution containing a number of other catalytic or promoter metal salts. This cogel procedure is similar to the gel preparative technique, but differs from prior methods based on the coprecipitation of two metal compounds together with a molecular sieve, and the subsequent addition, after dehydration, of a third or fourth metal component. The cogel procedure reportedly gives catalysts with unprecedented activity and activity maintenance for HDS and HDN of heavy feedstocks. In one case [40], the catalyst was treated at 1200°C for 2 h with no change in surface area (350 m^2 gm^{-1}). The stability is ascribed to the increased degree of structural homogeneity obtained by this technique. It has been used to prepare $NiO \cdot WO_3$ [40], $NiO \cdot WO_3 \cdot ZrO_2$ [42], and $NiO \cdot WO_3 \cdot TiO_2$ [41, 42] catalysts, all stabilized with ultrastable zeolites.

In the refractories area, some efforts have been made to prepare silicon carbide [44], silicon nitride [45], and other high-temperature compounds in forms applicable as supports. While the value of the materials so far produced over other more readily available supports is not apparent, some of their properties encourage further study. The high thermal conductivity of silicon carbide could be used advantageously in highly exothermic reactions, while aluminum borate (5–100 m^2 gm^{-1}, stable at 1300°C) [46] and boron phosphate (200 m^2 gm^{-1}, stable at 500°C) [47] may find use in the coal conversion area.

The last class of supports worth noting is that of monolith structures [48-51] which were developed primarily for automotive catalysts and are composed of small parallel channels of a variety of shapes and diameters. They may be in the form of honeycomb ceramics extruded in one piece, of oxidized aluminum alloys in rigid cellular configurations, or of multilayered ceramic corrugations. The channels in honeycomb ceramics are 1–3-mm tubes fabricated of low surface area mullite ($3Al_2O_3 \cdot 2SiO_2$) or cordierite ($2MgO \cdot 5SiO_2 \cdot 2Al_2O_3$). Recently silicon carbide, silicon nitride, and zirconia monoliths have become available. The refractory monoliths are macroporous (1–10 μm), and are usually coated with thin layers of catalytic materials, 5–20 wt% coatings being common. The two major advantages of monolithic supports for catalyst operations are high geometric surface area and low pressure drop during operation. In comparing monoliths with packed beds, it is not unusual to observe a pressure drop decrease of over one order of magnitude for the same geometric surface area. These advantages combined with the increased thermal conductivity of the monoliths make these structures support candidates for such highly exothermic reactions as methanation.

In the preparation of a supported metal catalyst, it is of primary importance to maximize metal dispersion and to maintain such dispersion during reaction in order to assure maximum catalytic activity and activity maintenance. Crystallite size and metal–support interactions can become noticeably significant at high dispersions. These will not be emphasized here, but are treated in detail in Chapter 3.

The choice of the proper procedure for maximizing dispersion of the metal on a support depends upon the chemistry involved and its compatability with the surface properties of the support. The most common technique is impregnation of the support with an aqueous solution containing salts of the catalytic metals. This is usually accomplished by incipient wetness [52]; i.e., using only enough solution to just fill the pores of the support. Sometimes the catalyst is evacuated prior to impregnation to insure homogeneous distribution of the solution throughout the catalyst pellet [53]. The impregnation technique has been successfully used to prepare bimetallic cluster [54, 55] and alloy [56] catalysts and can also be used in modified form to impregnate the surface of the pellets [57]. This is accomplished by first impregnating the pellets with an organic liquid (e.g., toluene) to such an extent that only a thin surface layer of the pellet is left dry. This is then impregnated with an aqueous solution of the metal salts to give a surface coated pellet. On drying, the solvent is driven off. This form of catalyst is useful for reactions which run under mass transfer limitations or can generate excessively high temperatures in the pellet center.

An interesting control of impregnation is provided by the displacement of immiscible liquid impregnation procedure [58]. It involves slurrying the support with a water-immiscible liquid in a high-speed blender and adding predetermined amounts of the aqueous impregnating solution. Because most common oxide supports such as alumina or silica are more hydrophilic than oleophilic, the aqueous solution displaces the water-immiscible liquid from the support pores and impregnation is accomplished to the desired extent. Preparative parameters are easily controlled for reproducibility.

Supported metal catalysts are also prepared in high dispersion by cationic [59] or anionic [59, 60] adsorption of the metal species from solution to the surface of the support. The crucial aspects of the surface chemistry of the support and its interaction with the solvent have only recently been recognized [60].

A technique yielding relatively stable and highly dispersed supported metal catalysts for application in coal conversion processes is the homogeneous precipitation [61–63]. In this method, the precipitating hydroxyl ions are generated throughout a suspension of the support in a solution of the active metal by the decomposition of urea whereby concentration gradients are avoided and a homogeneous precipitate with higher surface area is obtained. If the precipitating compound nucleates more easily at the surface of the carrier than in pure solution, a precipitation onto the support is obtained. This procedure produced SiO_2-supported catalysts containing 13–70 wt% of Ni having crystallite sizes of 15–40 Å [62, 63], equivalent to nickel surface areas of 170–450 m^2 gm^{-1} Ni. The technique has broad application, and should also find use in synthesizing multimetallic systems.

A few other techniques ensuring high dispersion deserve mentioning. The aerogel procedure [26–30] for the preparation of dispersed supports or supported metals has been discussed (see Section 5.2.2). The use of organic transition metal complexes with bonding between different metals is of particular value in preparing structurally well-defined multimetallic systems [64, 65]. Thus [64], a supported cobalt–rhodium catalyst was prepared by impregnating silica with a hexane solution of $[Co_2Rh](CO)_{12}$, followed by the decomposition and reduction. A strong interaction between the metals was indicated by the catalytic properties of the bimetallic system which were significantly different from those of the pure metals [64]. This preparative procedure has value in exploring the utility of model multimetallic systems in a number of areas (e.g., CO/H_2 synthesis reactions).

In the organometallic area, one also has to consider the use of such complexes both in the homogeneous phase and "anchored" to supports. These catalysts (for details, see Chapter 9) and preparative procedures are likely to find application in

CO/H_2 synthesis because of the possibility of improved selectivity at milder conditions, and increased resistance to sulfur poisoning [66]. On the other hand, because of problems of stability and catalyst recovery, such systems are inapplicable to direct coal liquefaction to boiler fuels [29].

Another important aspect of catalyst preparation is maintenance of metal dispersion. The formation of bimetallic alloys for such purpose [67] has already been discussed in Chapter 2, as has the use of metal support interactions [68, 69] in Chapter 3. There are two other procedures with broad applications that need mentioning.

In one, a trimetallic alloy or cluster is formed to stabilize two normally immiscible metals [70]. The primary requirement of the third metal is that it be miscible with both other metals. Thus, Ru–Cu "alloys" or clusters can be stabilized by the addition of nickel, which is miscible with both ruthenium and copper. In a similar manner a stable Ru–Pt–Cu catalyst has been prepared [70].

The second stabilization procedure is designed to prevent macroscopic migration of metal salt solutions during drying after impregnation. Such migration is caused by capillary forces in monolith supports impregnated with a platinum salt solution. It leads to a concentration gradient across the monolith and to uneven catalytic activity [71]. This problem can be eliminated by treating the impregnated monolith with H_2S and thereby converting the platinum salt to monodispersed insoluble sulfide [72] which cannot migrate during drying. A brief air calcination and subsequent hydrogen reduction gives a highly and uniformly dispersed metal catalyst. The technique has broad applications in the preparation of various metallic or multimetallic catalysts.

5.3 NONMETALLIC CATALYSTS

This category includes oxides, sulfides, and oxysulfides which have applications in coal conversion [29]. Many techniques used to prepare these materials are similar to those previously described for the preparation of metallic catalysts. In the following sections unsupported and supported catalysts will be discussed separately.

5.3.1 Unsupported Catalysts

High surface area, unsupported catalysts, or catalyst supports are accessible by methods previously discussed, such as flame decomposition [12], freeze-drying [14–18], sol–gel formation [23–25], gel precipitation [23–25, 73], homogeneous pH control [61, 62], and aerogel formation [26–28]. By controlling the various preparative variables, surface area, pore-size distribution, and mechanical strength can be varied within broad limits. Most of these procedures lead to oxides or mixed oxides. In some cases, sulfides or oxysulfides can be generated by sulfidation. Other chemical treatments can yield alternate materials as well. For example, combination of freeze-drying and high-temperature treatment of a colloidal graphite–tungstic acid solution produces high surface area tungsten carbide ($d \sim 350$ Å) [14]. Similarly, nitrides, borides, and phosphides with reasonable surface areas [29, 44–47] have been prepared.

A procedure worth noting is that of impregnation synthesis [74] which provides access to novel inorganic compounds normally prepared by high-temperature solid-state reactions in catalytically inactive low surface area form. In a particular example, it was desired to synthesize a series of materials with the general formula $M_2Mo_3O_8$ where M = Mg, Zn, Mn, Cd, Fe, Co, or Ni [74]. A structural feature of these materials is the presence of clusters of three Mo cations in which the Mo–Mo distance, 2.53 Å, is smaller than that in molybdenum metal, indicating strong bonding within these clusters. In order to prepare high surface area materials, high surface area MgO was impregnated with $(NH_4)_6Mo_7O_{24} \cdot 4H_2O$, followed by calcination in an H_2/H_2O environment at 500–700°C to induce a surface reaction. After removal of excess MgO with dilute HCl and drying, pure $Mg_2Mo_3O_8$ with a surface area of 100^2 gm^{-1} was obtained. Thus impregnation synthesis involves the use of one reactant as a high surface area template. In the subsequent reaction the formation of the surface compound is generally much faster than the corresponding solid-state reaction because of the high dispersion of the reactants and therefore occurs at significantly lower temperatures than the bulk reaction.

5.3.2 Supported Catalysts

Oxides of transition metals, such as chromium, cobalt, molybdenum, iron, and nickel are readily deposited on supports by all of the procedures previously described for the preparation of supported metal catalysts. The more common techniques include impregnation procedures [52, 53, 57], adsorption from solution [59, 60], and gel precipitation [23, 73]. For example, chromia supported on alumina is easily prepared by incipient wetness impregnation of alumina with an aqueous solution of a chromium salt, such as chromium nitrate or ammonium dichromate, or of chromic acid [75] followed by drying and calcination. Similar catalysts can also be obtained by precipitation procedures by adding a solution of chromium nitrate to a suspension of alumina in ammonium hydroxide [76]. Alternatively, chromia–alumina can be prepared by coprecipitation by the addition of ammonium hydroxide to an aqueous solution of aluminum and chromium nitrates. In general, the impregnation is useful for preparing weight loadings below 20 wt%, while, for loadings of 20–60 wt%, precipitation or gelation are more effective. Similar techniques are used for other transition oxides or mixed oxides.

A commercially widely used supported nonmetallic catalyst is cobalt molybdate on alumina. The description of the preparation of this catalyst using an alumina support with a controlled pore-size distribution [77] follows. Silica-stabilized alumina, with greater than 50% of its surface area in 30–80-Å pores and at least 3% of the total pore volume in pores greater than 2000 Å in diameter, was impregnated with an aqueous solution of salts of cobalt and molybdenum. The finished oxysulfide catalyst was tested for hydrodesulfurization of petroleum atmospheric residuum at 700°F and 1500 psig for 28 days and compared with a conventional cobalt molybdate catalyst having a major portion of the surface area in 30–70-Å pores. The latter catalyst retained 57% of its original activity, while the controlled pore catalyst maintained 80% activity.

Supported nonmetallic catalysts are more difficult to characterize than supported metallic catalysts because of lack of generally applicable selective chemisorption procedures for the determination of the specific area of the active species. While certain x-ray, magnetic, and spectroscopic techniques are available for the characterization of nonmetallic catalysts, the state of understanding of such systems is behind that for supported metal catalysts.

5.4 PRACTICAL IMPLICATIONS

The following is a summary of key points made in this chapter in connection with the catalytic conversion of coal.

5.4.1 Dispersed Materials

Numerous new compounds are synthesized by solid-state and inorganic chemists each year. Many of these materials contain elements of catalytic interest, either in preferred oxidation states or in unusual structural forms. A wealth of discovery awaits the catalytic scientist who can prepare such materials with sufficiently high surface area for characterization and catalytic testing. Techniques such as impregnation synthesis [74], flame decomposition [12], and freeze-drying [14–18] offer a means to fulfill this need for many systems. These and other preparative procedures mentioned in Table 5-1 are expected to find broad applications in coal conversion catalysis. Even if some of these techniques lack commercial utility, they may provide a means for screening new materials in the laboratory in a most expeditious way.

5.4.2 Supported Catalysts

The displacement of immiscible liquid procedure should be further explored since it can control the extent of deposition of the catalytic phase in pores of a given size range.

Better understanding of the reactions occurring at the liquid–solid interface during adsorption should lead to procedures for preparing very highly dispersed nickel, cobalt, iron, copper, silver, gold, and ruthenium. Such improved methods would have a significant impact on the preparation of multimetallic clusters. Applications of this work are expected in the areas of CO/H_2 synthesis and general refining and upgrading of coal liquids. The anchoring of organometallic complexes may find application in two areas: CO/H_2 synthesis (especially methanation and methanol synthesis), and possibly in water–gas shift catalysis. This expectation is based on the assumptions that organometallic complexes active in the reduction of CO with H_2 can be synthesized, and that such complexes will be sulfur tolerant.

5.4.3 Controlled Physical Property Supports

Advances in this area will be important for preparing support materials with controlled pore-size distributions, optimal surface acidity, and improved mechanical properties. The aerogel technique [26–28] has significant promise in this respect. Preparative variables are readily controlled, and the method is simple. This procedure is also applicable to the synthesis of mixed oxides, sulfides, and oxysulfides. Areas of impact are direct liquefaction, CO/H_2 synthesis, and water–gas shift catalysis. For direct liquefaction, controlled-pore supports have proven to be very useful in extending catalyst life [77]. Wide pores are necessary to prevent geometric blockage by inorganic material deposited on the surface, while small pores provide the surface area needed for high catalytic activity. Obviously, a proper balance is required between the two types of pores.

5.4.4 Stabilized Supports

Stabilized supports are needed for highly exothermic catalytic reactions (e.g., methanation, water–gas shift) and for processes involving oxidative regeneration of catalysts. The preparative procedures for thermally stabilizing aluminas and zeolites will have significant value in preparing new catalysts for direct liquefaction and CO/H_2 synthesis reactions. HDS, HDN, and hydrocracking catalysts stable at 1000°C already exist [39–43]. The impact in refining and upgrading is therefore apparent.

The engineering and materials science technology which has culminated in the development of refractory and monolithic supports may have application for gas-phase reactions involving heat and mass transfer problems.

5.4.5 Stabilized Supported Metals

While metals are not likely to be useful as direct liquefaction catalysts because of sulfidation, they are expected to be important in CO/H_2 synthesis reactions in stabilized form and as multimetallic systems for controlling product distribution and enhancing sulfur tolerance. New stabilization procedures such as the use of bimetallic [54, 55, 67] and trimetallic [70] systems could be very useful in this area and in upgrading and refining of coal liquids. The metal–support stabilization procedures developed for automotive catalysis [68, 69] are expected to be important for the preparation of thermally stable and sulfur-tolerant catalysts (see Chapter 3).

5.4.6 Multimetallic Catalysts

Many of the statements made in the above section apply to this area as well. Noteworthy work has been done with Raney metal catalysts in such applications as packed beds [20, 21], flame-sprayed coatings [19], and multimetallic catalysts [20].

REFERENCES

1. Sinfelt, J. H., *J. Catal.* **29**, 308 (1973).
2. Innes, W. B., *in* "Catalysis" (P. H. Emmett, ed.), Vol. 1, pp. 245–314. Van Nostrand-Reinhold, Princeton, New Jersey, 1954.
3. Ciapetta, F. G., and Plank, C. J., *in* "Catalysis" (P. H. Emmett, ed.), Vol. 1, pp. 315–352. Van Nostrand-Reinhold, Princeton, New Jersey, 1954.
4. Gil'debrand, E. I., *Int. Chem. Eng.* **6**, 449 (1966).
5. Sinfelt, J. H., and Cusumano, J. A., *in* "Advanced Materials in Catalysis" (J. J. Burton and R. L. Garten, eds.). Academic Press, New York, 1977.
6. *Int. Symp. Sci. Basis Catal. Preparation, Brussels* October 14–17, 1975.
7. Brown, H. C., and Brown, C. A., *J. Am. Chem. Soc.* **84**, 1493 (1962).
8. Brown, H. C., and Brown, C. A., *J. Am. Chem. Soc.* **85**, 1003 (1963).
9. Carter, J. L., Cusumano, J. A., and Sinfelt, J. H., *J. Catal.* **20**, 223 (1971).
10. Kullifay, S. M., *J. Am. Chem. Soc.* **83**, 4916 (1962).
11. Holt, E., *Nature (London)* **203**, 857 (1964).
12. Formenti, M., Juillet, F., Meriaudeau, P., Teichner. S. J., and Vergnon, P., *J. Colloid Interface Sci.* **39**, 79 (1972).
13. Bubidge, B. W., and Wood, D., British Patent No. 1,322,330 (1973).
14. Roehrig, F. K., and Wright, T. R., *J. Am. Ceram. Soc.* **55,** 58 (1972).
15. Tseung, A. C. C., and Bevan, H. L., *J. Mater. Sci.* **5**, 604 (1970).
16. Trambouze, Y., Applications de La Lyophilisation a la Preparation des Absorbants et Catalyseurs, Preprints for the *Int. Symp. Sci. Basis Catal. Preparation, Brussels, October* p. 13 (1975).
17. Roehrig, F. K., and Wright, T. R., *J. Vac. Sci. Technol.* **9**, 1368 (1972).
18. Meissier, D. R., Marblehead, and Gassa, G. E., U. S. Patent No. 3, 681, 010 (1972).
19. Haynes, W. P., Forney, A. J., Elliott, J. J., and Pennline, H. W., *Am. Chem. Soc., Div. Fuel Pet. Chem. Prepr.* **14**, 126 (1970).
20. Erzhanova, M. S., Beisekhov, T., Sokol'skii, D. V., and Khisametdinov, A. M., *Khim. Khim. Technol.* **72**, 71 (1973).
21. Yakubenok, E. F., and Podvyazkin, Yu. A., *Khim. Khim. Technol. Polim. Org. Sint.* **73**, 29 (1974).

22. Raney, M., *Ind. Eng. Chem.* **32**, 1199 (1940).
23. Dell, R. M., *in* "Reactivity of Solids" (J. S. Anderson, *et al.*, eds.), p. 553. Chapman and Hall, London 1972.
24. Fletcher, J. M., and Hardy, C. J., *Chem. and Ind.* **75** (1967).
25. Woodhead, J. L., *Sci. Ceram.* **4**, 105 (1968).
26. Astier, M. *et al.*, Preparation and Catalytic Properties of Supported Metal or Metal-Oxide on Inorganic Oxide Aerogels, *Symp. Sci. Basis Catal. Preparation, Brussels* October 1975.
27. Nicolaon, G. A., and Teichner, S. J., *Bull. Soc. Chim. Fr.* 1906 (1968).
28. Gardes, G. E. E., Nicolaon, G. A., Vicarini, M. A., and Teichner, S. J., *J. Colloid Interface Sci.* **5,** 245 (1976).
29. Boudart, M., Cusumano, J. A., and Levy, R. B., New Catalytic Materials for the Liquefaction of Coal, Rep. RP-415-1. Electric Power Res. Inst., October 30, 1975.
30. Hindin, S. G., and Pond, G. R., German Patent 2,458,221 (1975).
31. Hindin, S. G., and Pond, G. R., German Patent 2,458,122 (1975).
32. Gould, R. F., "Molecular Sieve Zeolites," Vol. 1. American Chemical Society, Washington, D.C., 1971.
33. Meier, W. M., and Uytterhoeven, J. B., (eds.), "Molecular Sieves." American Chemical Society, Washington, D.C., 1973.
34. Kerr, G. T., *J. Phys. Chem.* **73**, 2780 (1969).
35. Kerr, G. T., *J. Phys. Chem.* **71**, 4155 (1967).
36. Kerr, G. T., in "Molecular Sieves" (W. M. Meier and J. B. Uytterhoeven, eds.), p. 219. American Chemical Society, Washington, D.C., 1973.
37. Ward, J. W., U. S. Patent No. 3,781, 199 (1974).
38. Ward, J. W., U. S. Patent No. 3,897,327 (1975).
39. Kittrell, J. R., U. S. Patent No. 3,639,271 (1972).
40. Kittrell, J. R., U. S. Patent No. 3,558,471 (1971).
41. Kittrell, J. R., U. S. Patent No. 3,536,606 (1970).
42. Kittrell, J. R., U. S. Patent No. 3,536,605. (1970).
43. Kittrell, J. R., U. S. Patent No. 3,535,227 (1970).
44. Exell, S. F., Roggen, R., Gillot, J., and Lux, B., The Preparation and Structure of Ultrafine Carbide, *Int. Conf. Solid Compounds Transition Elements, 4th* p. 39 (1973).
45. Exell, S. F., Roggen, R., Gillot, J., and Lux, B., Preparation of Ultrafine Powders of Refractory Carbides in an ARC-Plasma, Fine Part., *Int. Conf., 2nd, Boston, October 7–11, 1973* p. 165 (1974).
46. McArthur, D. P., U. S. Patent No. 3,856,705 (1974).
47. Moffat, J. B., and Neeleman, J. F., *J. Catal.* **31**, 274 (1973).
48. Bagley, R. D., Doman, R. C., Duke, D. A., and McNally, R. N., SAE Preprints, Detroit, Michigan, Jan. 8–12, 1973.
49. Howitt, J. S., SAE Preprints, Detroit, Michigan, Feb. 25–Mar. 1, 1974.
50. Andersen, H. C., Romeo, P. L., and Green, W. J., *Engelhard Ind. Tech. Bull.* **8**, 100 (1966).
51. Campbell, L. E., Comparison of Monolithic Versus Particulate Reactor Designs For the Cracking of Cumene and the Dehydration of Ethanol, presented at the *Spring Symp. Phil. Catal. Club* May 7, 1974.
52. Sinfelt, J. H., *Annu. Rev. Mater. Sci.* **2,** 641 (1972).
53. Mehrotra, R. P., *Pet. Hydrocarbons Chem. Age (India)* **5**, 56 (1970).
54. Sinfelt, J. H., *J. Catal.* **29**, 308 (1973).
55. Robertson, S. D., Kloet, S. C., and Sachtler, W. M. H., *J. Catal.* **39**, 234 (1975).
56. Inami, S. H., and Wise, H., *J. Catal.* **26**, 92 (1972).
57. Higginson, G. W., *Chem. Eng. News* **81** (20), 98 (1974).
58. Tauster, S. J., *J. Catal.* **18**, 358 (1970).
59. Benesi, H. A., Curtis, R. M., and Studer, H. P., *J. Catal.* **10**, 328 (1968).

60. Brunelle, J. P., and Sugier, A., *C. R. Acad. Sci. Paris Ser. C* **276**, 1545 (1973).
61. Cartwright, P. F. S., Newman, E. J., and Wilson, D. W., *The Analyst* **92**, 663 (1967).
62. Geus, J. W., Dutch Patent Application 6,705,259 (1970).
63. Geus, J. W., Dutch Patent Application 6,813,236 (1971).
64. Anderson, J. R., and Mainwaring, D. E., *J. Catal.* **35**, 162 (1974).
65. Smith, G. C., Chojnacki, J. P., Dasgupta, S. R., Iwatate, K., and Watters, K., *Inorg. Chem.* **14**, 1419 (1975).
66. Berenblyum, A. S., Ronzhin, L. K., Ermolaev, M. V., Kalechits, I. V., Khidekel, M. L., *Izv. Akad. Nauk. SSSR, Ser. Khim.* **11**, 2650 (1973); *Chem. Abstr.* **80**, 95402w (1974).
67. Myers, J. W., and Prange, F. A., U. S. Patent No. 2,911,357 (1959).
68. Bernstein, L. S., NOx Reduction With Nickel-Copper Alloy and Stabilized Ruthenium Catalysts, preprints, *Jpn.-U.S.A. Sem. Catal. NOx Reactions, Susono, Japan* p. 8–1, November 2–4, 1975.
69. Dalla Betta, R. A., Gandhi, H. S., Kummer, J. T., and Shelef, M., U. S. Patent No. 3,819,536 (1974).
70. Sinfelt, J. H., and Cusumano, J. A., U. S. Patent No, 3,901,827 (1975).
71. Kobylinski, T. P. *et al.,* U. S. Patent No. 3,784,675 (1974).
72. Kobylinski, T. P., and Taylor, B. W., U. S. Patent No. 3,840,389 (1974).
73. Grimes, J. H., and Scott, K. T. B., *Powder Metall.* **11**, 213 (1968).
74. Tauster, S. J., *J. Catal.* **26**, 487 (1972).
75. Ciapetta, F. G., and Plank, C. J., *in* "Catalysis" (P. H. Emmett, ed.), Vol. 1, p. 343. Van Nostrand-Reinhold, Princeton, New Jersey, 1954.
76. Eischens, R. P., and Selwood, P. W., *J. Am. Chem. Soc.* **69**, 2698 (1947).
77. Riley, K. L., and Sawyer, W. H., U. S. Patent No. 3,770,617 (1973).

Chapter 6

Poisoning and Regeneration

6.1 INTRODUCTION

The prevention of catalyst poisoning and the necessity for catalyst regeneration present a number of severe and important constraints for the technical and economic feasibility of most coal conversion processes. These problems are of particular significance for the production of methane or liquids from coal, and may determine the type of the processes used. They parallel the poisoning and regeneration problems encountered in the upgrading of heavy petroleum crudes and residua. Thus for feed stocks with high sulfur and metal content the residuum desulfurization catalyst has a life of less than 30 days and can not be regenerated successfully. The need for frequently replacing the relatively expensive catalyst seriously affects the economics of the process. Similar poisoning problems may be encountered in certain coal liquefaction processes, and is definitely an important aspect in methanation economics.

In spite of the importance of catalyst poisoning and regeneration in coal conversion, only limited progress has been made in the last decade. The present discussion will highlight the lack of attention on the part of the scientific community and point out promising directions for research.

Catalyst deactivation can occur by fouling and poisoning. Fouling is the result of deactivation due to species that result from transformations of the feed. Carbonaceous residues are a typical example. Poisons, on the other hand, are not originally a chemical part of the reacting material. They are impurities introduced in the feed stream. Examples are H_2S in methanation and Fischer–Tropsch synthesis, CO and mineral matter in coal liquefaction.

In the processes of interest in coal conversion, the most severe poisons and fouling agents are sulfur, carbon residues, and inorganic materials. In liquefaction, activity decay is caused by the accumulation of carbon on the catalyst as well as the deposition of metal compounds and minerals on the catalyst surface. In the upgrading of coal liquids, carbon deposition again is a primary problem, as is the severe poisoning effect of sulfur. Methanation catalysts are particularly sensitive to H_2S, and in current processes the H_2S content must be kept below 1 ppm. The same holds true for Fisher–Tropsch catalysts, where H_2S, carbon residues, and chlorine cause catalyst deactivation.

The main advance in this area has been in the modeling of deactivation phenomena [1–11] with particular emphasis on coke deposition.

6.2 SULFUR

The nature of the effect of sulfur compounds on the catalyst depends on their concentration. High concentration of sulfur compounds can affect the bulk composition of the catalyst and may lead to complete sulfidation. When present at low concentration, the sulfur species only competes with the reactants for adsorption sites. Intermediate concentrations of sulfur can cause a restructuring of the catalytic surface, changing the activity to a much larger degree than expected from simple competition for active sites. The extent of any of these processes is a function of the thermodynamics of the catalytic material in the reactive environment. Only few thermodynamics studies of this type have been reported [12, 13]. Bulk considerations have been recently included in a review of catalytic materials of interest to coal liquefaction [13]. It was found that, at the high H_2S levels encountered in liquefaction processes (over 1%), most compounds can sulfide. Exceptions are the oxides Al_2O_3, SiO_2, and TiO_2, and certain borides, phosphides, and silicides of transition metals. The sulfiding tendency decreases with decreasing severity of the environment. Platinum, for example, is expected to resist bulk sulfidation around 400°C at a level of less than 100 ppm H_2S in H_2. Above this concentration interaction of H_2S with the metal is expected. In the case of Ni at 550°C, a monolayer of sulfide forms at 2–5 ppm H_2S in H_2 [14], while on iron catalysts at 400°C this occurs at 1.6 ppm H_2S [15]. On the latter catalyst only a fraction of the monolayer coverage is required to inhibit ammonia synthesis completely.

In many other instances it has also been observed that sulfur adsorption on a catalyst leads to a much more pronounced loss in activity than expected from the loss of active surface area [16]. Somorjai [17] suggests that this effect is due to a reconstruction of the surface of the catalyst—a consequence of the difference in surface energy of the various low-index planes of a solid. According to this concept, adsorption of small amounts of H_2S modifies the surface energy balance and leads to a new equilibrium distribution of surface planes with different catalytic activity. This explanation would lead to the conclusion that structure sensitive reactions would be affected to a much greater degree by this type of poisoning than would structure insensitive reactions. Such an effect has been actually reported by Maurel *et al.* [18], who find a striking difference in the effect of sulfur poisoning on two Pt-catalyzed reactions: the hydrogenolysis of cyclopentane and the hydrogenation of benzene. The difference was observed only when H_2S and SO_2 were introduced together and the tendency to produce elemental sulfur was the highest.

The complexity of the sulfur-poisoning problem becomes clearly apparent when more than one component is present in a catalyst system, since the effect of sulfur will not be uniform. Steam reforming catalysts, for example, usually contain promoters in addition to nickel. Some of the promoters bind sulfur more strongly than others. Thus, Rostrup-Nielsen reports that generally Na and K form sulfates that cannot be regenerated with steam at 600°C [14]. The unpromoted catalyst, on the other hand, and catalysts promoted with Mg and Ca, regenerate easily at these conditions [14]. In CuO–ZnO shift catalysts, both components have been found to sulfide in H_2S [19], and the activity to correlate with CuO area.

Studies have been made to exploit the reactivity differences of two catalyst components to enhance sulfur resistance in several applications. The addition of Cr_2O_3 to nickel–zeolite catalysts has been reported to increase sulfur tolerance for ethane hydrogenolysis [20]. This is an effect that should be explored further.

6.2.1 Kinetics and Mechanism

Quantitative studies of the kinetics of H_2S adsorption on single-crystal surfaces [25, 26] reveal that there are two adsorption regimes for Pt and Cu. At coverages below 0.5–0.6 of a monolayer, the adsorption is fast with a high sticking coefficient. Above this coverage, the adsorption process slows down considerably, and has a very low sticking coefficient. The nature of this phenomenon is not understood, but it is clear from these observations that the adsorption of H_2S on a metal surface affects its behavior more severely than expected from mere one-to-one interaction of a S atom with a surface atom. Surface reconstruction, discussed earlier, may be only one of the possible explanations. An interruption of the continuity of the metal surface may also be an important factor in the behavior of a poisoned catalyst, as well as the possibility that adsorption on one site affects the neighboring sites. In their methanation study [27] Dalla Betta *et al.* made observations that point in this direction and have important practical consequences. They find a strong effect of H_2S poisoning on the selectivity of a number of catalysts, including Ni and Rh on various supports, and speculate that the effect is due to the type of site interruption mentioned above.

Interest in sulfur-poisoning models and mechanisms has been limited as most of the fouling or poisoning models in the literature have been applied primarily to carbon deposition. However, Carberry and Gorring [7] showed that the models also apply to sulfur poisoning in certain cases. Thus the observations of Anderson *et al.* [21] on the increasing sulfur resistance of iron Fischer–Tropsch catalyst with decreasing pellet size and the presence of sulfur in a thin layer near the catalyst surface is consistent with the shell-progressive mechanism developed for carbon deposition on catalyst pellets [7].

6.2.2 Regeneration

The question of regeneration of sulfur-poisoned catalysts has been discussed by several investigators [14, 15, 17, 25, 28]. Rostrup-Nielson [14] found that the ease of regeneration of sulfur-poisoned nickel catalysts varies with the promoters used in the catalyst system, promoters that bind sulfur most strongly causing the most difficult regeneration. Somorjai [17] makes an interesting suggestion about promoters that tend to improve poison resistance [22–24]. After speculating that small amounts of sulfur cause surface reconstruction of catalysts such as Pt and Ni because they lower the surface free energy of certain crystal faces, he concluded that the addition of compounds that would have the opposite effect on surface free energy would either prevent or reverse surface reconstruction and thus enhance sulfur resistance. A patent [28] on this subject is not very explicit.

6.2.3 Activity Maintenance

While the earlier discussion stressed the adverse effects of sulfur poisoning, there is one application that makes use of the strong adsorption of H_2S. In catalytic reforming, H_2S is purposely added to a stream to moderate the activity of a Pt or a bimetallic catalyst in order to minimize initial coke deposition [29, 30]. The function of H_2S is to minimize hydrogenolysis activity (cracking) which leads to high surface concentrations of unsaturated carbon species. The latter rapidly ''polymerize'' to coke and foul the catalyst.

6.3 CARBON

Catalyst deactivation by carbon deposition occurs in most hydrocarbon reactions. To minimize coking, catalytic reforming is carried out under high hydrogen pressure in spite of its adverse effect on the kinetics of some of the desired reactions. In methanation, carbide formation ranks as one of the most severe problems, second only to sulfur poisoning. In hydrodesulfurization, catalyst life is severely limited by heavy carbon deposition, in particular with heavy crudes and residua.

In spite of the universal nature of the problem, only limited attention has been given to several key questions concerning carbon deposition. In one of the few studies on this subject Eberly and co-workers [1] investigated the nature of coke on cracking catalysts by infrared spectroscopy, and concluded that the carbon is present in the form of highly condensed, aromatic structures of low hydrogen content. The situation with coal conversion catalysts is likely to be more complex. In hydrodesulfurization two distinct deactivation regions are observed [2], and it is likely that the nature of the carbon containing residues are different in these two regions. The understanding of this phenomenon would benefit from investigations of the type conducted by Eberly.

6.3.1 Modeling of Deactivation

Most of the work on coke deposition deals with modeling of the deposition reaction [1–11]. Most models use empirical power laws applicable only in limited space velocity regions because the coking rate is a function of space velocity [1]. In expressions of the type

$$C = At^n \tag{6-1}$$

where C is the weight percent carbon on the catalyst, t is time, and A and n are constants; the value of n can vary from 0.4 to 1.0 in the 0.2–30 V/V/h space velocity range [1]. A diffusion-controlled expression ($n = ½$) is only expected in a small region of operation. In spite of the variation in n, expression (6-1) has been successful in modeling several systems, in particular zeolite and SiO_2/Al_2O_3 deactivation in cracking reactions [2, 4, 31].

Other expressions have been used to describe fouling in cracking reactions. The simplest is the first-order decay model

$$da/dt = Aa \tag{6-2}$$

where a is the rate of reaction normalized to the initial (unpoisoned) rate [4, 31].

Efforts to derive general poisoning equations and to relate parameters to specific deactivation mechanisms have intensified in the last decade [5–7, 9–11]. They usually incorporate parallel reactions (fouling due to side reactions of the main reactants), series reactions (fouling due to further reaction of the primary product), and poisoning by species not derived from the reactants.

Equation (6-2) is in fact a limiting case of a general expression, derived by

Khang and Levenspiel [5], which incorporates these three types of deactivation mechanisms:

$$da/dt = Aa^d \tag{6-3}$$

In this expression, the constant d varies from 1 to 3, depending on the controlling mechanism. The value $d = 1$ is obtained for several models, including uniform deactivation (i.e., no diffusion resistance) for both parallel and series reactions [5]. The inclusion of diffusion limitations in these models can have a very important effect on the results. Thus, in certain cases discussed by Lee and Butt [11], the presence of some diffusion limitation results in increased catalyst life.

One diffusion model of interest is the shell-progressive model, where the reactants have to diffuse through a growing nonreactive shell of the catalyst pellet [6, 7]. This effect is expected in cases in which diffusion resistance is large compared to the fouling rate, and is found to apply most successfully for parallel fouling reactions [6]. General relationships for this type of mechanism have been developed [7]. The specific determination of the controlling fouling mechanism has been explored experimentally and theoretically by Hegedus and Petersen [32–35]. By using a single pellet reactor and monitoring the reactant concentration in the exterior and center plane of the pellet, the authors were able to determine many of the rate parameters of the reaction. Recently their computations have been extended from first- to higher-order reaction rates [10]. The order of the reaction affects the relative rates mostly in parallel poisoning reactions, but has a minor effect on pore-mouth poisoning models, and on poisoning by impurities.

6.3.2 Regeneration

The regeneration of coked catalyst in coal conversion reactions has received only minor attention, indicating the complexity of the problem. In the petroleum industry, the regeneration of coked catalysts is commonly achieved by controlled slow oxidation. In the absence of diffusion effects, this regeneration has been found to be first order in O_2 and C [36]. The kinetics of carbon combustion under completely diffusion-controlled conditions is more complex [37]. In coal conversion, catalysts used with feeds with low ash, metal, and mineral content can be regenerated by oxidation [38, 39], but at higher concentrations these impurities cause irreversible complications during oxidative regeneration [38]. In case of metal poisons, the regeneration procedures suggested involve the complete extraction of the metal catalyst from the support [40].

6.4 OTHER POISONS

In addition to carbon and sulfur, the poisons of most concern in petroleum and coal processing are mineral matter (salts such as NaCl) and metals such as V, Ni, and Ti [39]. In certain reactions that require acidic catalysts (zeolites or SiO_2/Al_2O_3 catalysts), nitrogen compounds also present a serious problem [41, 42].

The literature is limited to confirmation of the poisoning effect of metals and alkali salts [38, 39]. Little is known concerning the specific poisoning mechanism or the nature of the poisons on the catalyst surface.

6.5 CONCLUDING REMARKS

The development of effective catalyst regeneration schemes is expected to become more critical as the importance of coal conversion and upgrading increases. It will become an important economic consideration in the choice of a conversion system. The current understanding of the effect of most fouling agents and poisons is limited. An increased effort is needed for the development of improved regeneration procedures for HDS catalysts poisoned by the alkali, vanadium, and titanium contaminants. The nature of these poisons has to be established, and the real extent of their interference with traditional regeneration procedures clarified. A similar effort is required to develop sulfur tolerance in methanation catalysts if this route of coal conversion is to be economically viable.

In particular, there is a need for a better understanding of the interaction of poisons and fouling agents with the catalyst and the support. The reasons for the large changes in surface area upon oxidative regeneration of Na and V containing HDS catalysts, for example, may be linked with solid-state reactions which could be inhibited or avoided if their nature could be clarified. The efforts to develop sulfur-tolerant catalysts would also benefit considerably from an understanding of the solid-state and surface chemistry of the sulfur–catalyst systems. Similarly, sulfur-induced surface reconstruction and selectivity changes should be explored further since they may have important implications on processes such as methanation and Fischer–Tropsch synthesis.

REFERENCES

1. Eberley, P. E., Jr., Kimberlin, C. N., Miller, W. H., and Drushel, H. V., *Ind. Eng. Chem., Proc. Res. Develop.* **5**, 193 (1966).
2. Ozawa, Y., Bischoff, K. B., *Ind. Eng. Chem., Proc. Res. Develop* **7**, 67 (1968).
3. Levinter, M. E., Panchekov, G. M., and Tanatarov, M. A., *Int. Chem. Eng.* **7**, 23 (1967).
4. Weekman, V. W., Jr., and Nace, D. M., *AIChE J.* **16**, 397 (1970).
5. Khang, S. J., and Levenspiel, O., *Ind. Eng. Chem., Fundamentals* **12**, 185 (1973).
6. Masamune, S., and Smith, J. M., *AIChE J.* **12**, 384 (1966).
7. Carberry, J. J., and Gorring, R. L., *J. Catal.* **5**, 529 (1966).
8. Buyanov, R. A., and Afanas'ev, A. D., *Kinet. Katal.* **16**, 802 (1975).
9. Hegedus, L. L., and Peterson, E. E., *Chem. Eng. Sci.* **28**, 345 (1973).
10. Wolf, E., and Petersen, E. E., *Chem. Eng. Sci.* **29**, 1500 (1974).

11. Lee, J. W., and Butt, J. B., *Chem. Eng. J. (Lausanne)* **6**, 111 (1973).
12. Foroulis, Z. A. (ed.), "High Temperature Metallic Corrosion of Sulfur and Its Compounds." The Electrochem. Society, Corr. Div., New York, 1970.
13. Boudart, M., Cusumano, J. A., and Levy, R. B., New Catalytic Materials for the Liquefaction of Coal. Electric Power Res. Inst. Rep. RP415-1, October 30, 1975.
14. Rostrup-Nielsen, J. R., *J. Catal.* **21**, 171 (1971).
15. Brill, Von R., Schaefer, H., and Zimmerman, G., *Ber. Bunsenges. Phys. Chem.* **72**, 1218 (1968).
16. Smith, R. L., Naro, P. A., and Silvestri, A. J., *J. Catal.* **20**, 359 (1971).
17. Somorjai, G. A., *J. Catal.* **27**, 453 (1972).
18. Maurel, R., Leclercq, G., and Barbier, J., *J. Catal.* **37**, 324 (1975).
19. Ray, N., Rastogi, V. K., Mahapatra, H., and Sen, S. P., *J. Res. Inst. Catal., Hokkaido Univ.* **21**, 187 (1973).
20. Lawson, J. O., and Rase, H. F., *Ind. Eng. Chem., Prod. Res. Develop.* **9,** 317 (1970).
21. Anderson, R. B., Karn, F. S., and Shulte, J. F., *J. Catal.* **4**, 56 (1965).
22. Zubova, I. E., Rabina, P. D., Pavlova, N. Z., Kuznetsov, L. D., and Chudinov, M. G., *Kinet. Katal.* **15**, 1261 (1974).
23. Garlet, R., Paul, C., and Grandet, L., *C. R. Congr. Ind. Gaz.* **83**, 264 (1966).
24. Cox, J. L., Sealock, J. J., Jr., and Hoodmaker, F. C., American Chemical Society, Div. Fuel Chem., Preprint 19.
25. Bonzel, H. P., *Surface Sci.* **27**, 387 (1971).
26. Bonzel, H. P., and Ku, R., *J. Chem. Phys.* **58**, 4617 (1973).
27. Dalla Betta, R. A., Piken, A. G., and Shelef, M., *J. Catal.* **35,** 54 (1974).
28. Pieters, W. J., Freel, J., and Anderson, R. B.,U. S. Patent No. 3,674,707 (1972).
29. Sivasanker, S., Ramaswamy, A. V., *J. Catal.* **37**, 553 (1975).
30. Hayes, J. C., Mitsche, R. T., Pollitzer, E. L., and Homeier, E. H., Presentation before the Division of Petroleum Chemistry, American Chemical Society, Los Angeles, March 31–April 5, 1974.
31. Weekman, V. W., Jr., *Ind. Eng. Chem. Proc. Res. Develop* **7,** 90 (1968); and **8,** 385 (1969).
32. Hegedus, L. L., and Petersen, E. E., *Chem. Eng. Sci.* **28**, 69 (1973).
33. Hegedus, L. L., and Petersen, E. E., *Chem. Eng. Sci.* **28**, 345 (1973).
34. Hegedus, L. L., and Petersen, E. E., *J. Catal.* **28**, 150 (1973).
35. Hegedus, L. L., and Petersen, E. E., *Ind. Eng. Chem., Fundamentals* **11**, 579 (1972).
36. Weisz, P. B., and Goodwin, R. B., *J. Catal.* **6**, 227 (1966).
37. Weisz, P. B., and Goodwin, R. D., *J. Catal.* **2**, 397 (1963).
38. McColgan, E. C., and Parsons, B. I., Hydrocracking of Residual Oils and Tar. 6. Catalyst Deactivation by Coke and Metals Deposition, Can., Mines Branch, Res. Rep. R273 (1974).
39. Kovach, S. M., and Bennett, J., *Am. Chem. Soc. Div. Fuel Chem. Preprint* **20**, 243 (1975).
40. Mitchell, D. S., Rafael, S., Bridge, A. G., Cerrito, E., and Jaffe, J., U. S. Patent No. 3,791,989 (1974).
41. Jacobs, P. A., and Heylen, C. F., *J. Catal.* **34**, 267 (1974).
42. Voltz, S. E., Nace, D. M., Jacob, S. M., and Weekman, W. M., Jr., *Ind. Eng. Chem., Proc. Res. Develop* **11**, 261 (1972).

Chapter 7

Mechanism and Surface Chemistry

In presenting a summary of the generally accepted mechanisms of the various reactions involved in the catalystic conversion of coal, when possible, emphasis will be placed on the relation between catalytic properties and the course of a given mechanism. This will serve as an introduction to the more detailed treatment of these phenomena for actual conversion processes as discussed in Part III.

The present discussion is divided into six broad areas which encompass the key reactions involved in the processes discussed in Chapter 1:

CO/H_2 synthesis	Hydrodesulfurization
Water–gas shift	Polynuclear aromatic cracking
Hydrodenitrogenation	Carbon/coal gasification

7.1 CO/H_2 SYNTHESIS

7.1.1 General Comments

The catalytic synthesis of hydrocarbons from CO/H_2 mixtures has been known since Sabatier's work in 1902 and especially since the developments by Fischer and Tropsch during the 1920–1935 period. Following World War II, this effect extended to the United States where a significant amount of work was carried out by a number of industrial and government laboratories. Much of this work was de-emphasized or phased out in the late 1950s because of the ready availability of petroleum. There are a number of excellent reviews on this early work [1–7].

The marked increase in demand for petroleum over the last five years has generated renewed interest in the CO/H_2 synthesis area as a means of producing both fuels and chemicals. The general synthesis reactions of interest are:

$$\text{paraffins} \quad (2n+1)H_2 + nCO \longrightarrow C_nH_{2n+2} + nH_2O \tag{7-1}$$

$$\text{olefins} \quad 2nH_2 + nCO \longrightarrow C_nH_{2n} + nH_2O \tag{7-2}$$

$$\text{alcohols} \quad 2nH_2 + nCO \longrightarrow C_nH_{2n+1}OH + (n-1)H_2O \tag{7-3}$$

These reactions may be accompanied by one or more of the following reactions:

$$\text{water-gas shift} \quad CO + H_2O \longrightarrow CO_2 + H_2 \tag{7-4}$$

$$\text{Boudouard reaction} \quad 2CO \longrightarrow C + CO_2 \tag{7-5}$$

$$\text{coke deposition} \quad H_2 + CO \longrightarrow C + H_2O \tag{7-6}$$

which sometimes can lead to difficulties in a kinetic investigation.

Most of the early research in this area was done before adequately sensitive analytical instruments became available and detailed product distributions could be determined. High conversion levels were required to obtain adequate samples for analysis, and therefore few kinetic studies were conducted in differential reactors or at conditions assuring reasonably good heat and mass transfer.

In the early work done before the development of chemisorption techniques for measuring metal surface areas, specific rates per unit surface area were not available. As shown by Vannice [8, 9], there are significant differences in the relative activities for the Group VIII metals when measured on a specific area and the early data reported for these metals by Fischer and co-workers [10]. A particularly striking finding is that iron is actually a very active methanation catalyst, and that the lower activity reported by Fischer was probably due to low surface area (see Section 4.5).

The developments in catalysis over the last decade are expected to contribute materially to the understanding of the mechanisms of complex surface reactions occurring in the CO/H_2 synthesis. The following discussion in this area covers methanation and the synthesis of liquid hydrocarbons.

7.1.2 Methanation

The methanation reaction is the simplest of the paraffin synthesis reactions [Eq. (7–1)]:

$$3H_2 + CO \longrightarrow CH_4 + H_2O \tag{7-7}$$

Methane is also formed by two other reactions:

$$2\,H_2 + 2\,CO \longrightarrow CH_4 + CO_2 \qquad (7\text{-}8)$$

$$4\,H_2 + CO_2 \longrightarrow CH_4 + 2\,H_2O \qquad (7\text{-}9)$$

but hydrogenation of CO_2 (7-9) does not occur in the presence of CO [11, 12], and eq. (7-8) can be considered to be a combination of eqs. (7-7) and (7-4). Much of the recent kinetic work in this area has been discussed in detail elsewhere [8–19]. The two mechanisms which appear to be consistent with the reported experimental data are those suggested by Vlasenko and Yuzefovich [20] and by Storch *et al.* [6]. Both of these groups postulate the formation of an adsorbed HCOH complex in the enol form:

$$\mathrm{H{-}\overset{\displaystyle OH}{C}{=}M}$$

which has also been proposed by other workers for the CO/H_2 reaction over Ru [21], Pt [22], Ni [20, 23], Co [24], and Fe [25–27], and for all the Group VIII metals [9]. The Russian workers suggest the following scheme:

$$* + e + H_2 \longrightarrow \underset{|\,*}{H_2^-} \qquad (7\text{-}10)$$

$$* + CO \longrightarrow \underset{|\,*}{CO^+} + e \qquad (7\text{-}11)$$

$$\underset{|\,*}{CO^+} + \underset{|\,*}{H_2^-} \longrightarrow \underset{|\,*}{HCOH^+} + e + * \qquad (7\text{-}12)$$

$$\underset{|\,*}{HCOH^+} + \underset{|\,*}{H_2^-} \longrightarrow \underset{|\,*}{CH_2} + H_2O + * \qquad (7\text{-}13)$$

$$\underset{|\,*}{CH_2} + \underset{|\,*}{H_2^-} \longrightarrow CH_4 + e + 2* \qquad (7\text{-}14)$$

where reaction (7-12) is the rate-determining step. These authors suggest that CH_2 polymerization predominates on metals which give higher molecular weight products [20].

An alternate mechanism has been derived [11] from the more general equations for liquids synthesis proposed by Storch *et al.* [6]:

$$\mathrm{O{=}C{=}M} \xrightarrow{2\,H} \mathrm{H{-}\overset{OH}{C}{=}M} \xrightarrow{H} \mathrm{H_2\overset{OH}{C}{-}M} \xrightarrow[-H_2O]{2\,H} \mathrm{H_3C{-}M} \xrightarrow{H} CH_4 \qquad (7\text{-}15)$$

In a recent study Vannice has investigated the kinetics of the methanation reaction over all of the Group VIII metals except osmium [8, 9], and explained his data by a kinetic model similar to that proposed by the Bureau of Mine workers [6, 28]:

$$\mathrm{CO} \rightleftharpoons \underset{*}{\underset{|}{\mathrm{CO}}} \tag{7-16}$$

$$\mathrm{H_2} \rightleftharpoons \underset{*}{\underset{|}{\mathrm{H_2}}} \tag{7-17}$$

$$\underset{*}{\underset{|}{\mathrm{CO}}} + \underset{*}{\underset{|}{\mathrm{H_2}}} \rightleftharpoons \underset{*}{\underset{|}{\mathrm{HCOH}}} + * \tag{7-18}$$

$$\underset{*}{\underset{|}{\mathrm{HCOH}}} + \frac{y}{2}\underset{*}{\underset{|}{\mathrm{H_2}}} \xrightarrow{\mathrm{RDS}} \underset{*}{\underset{|}{\mathrm{CH}_y}} + \mathrm{H_2O} \tag{7-19}$$

$$\underset{*}{\underset{|}{\mathrm{CH}_y}} + \mathrm{H_2} \xrightarrow{\text{fast}} \mathrm{CH_4} \tag{7-20}$$

In this work the number of H atoms, y, interacting in the rate-determining step (7-19) was found to increase with increasing ability of the metal to produce higher molecular weight species:

Metal	y
Pd	1
Ni	2
Co	3
Ru	4

Ru exhibits a marked tendency to form high molecular weight paraffinic waxes at high pressures and low temperatures, and was found to yield in this kinetic study carried out at atmospheric pressure the largest C_{5+} molecular distribution. This suggests that such studies at ambient pressures and at differential conditions may indicate the catalytic properties the metals would show at high conversions and pressures. Interestingly, it was also found that activity for methanation increased with H–metal bond strength and decreased with the CO–metal bond strength. This is consistent with the assumption that the bond strength of the metal–HCOH surface complex is directly related to the CO–metal bond strength. As the latter decreases, the surface coverage of the complex is also expected to decrease and make more of the surface available for hydrogen. This tends to equalize the surface coverages of HCOH complex and H_2, and thereby accelerates the rate-determining step [RDS (7-19)]. While the existence of the postulated enol intermediate has not been proven, the kinetic model involving it offers a reasonable explanation for the data obtained with all of the Group VIII metals.

7.1.3 Liquid Hydrocarbons

The formation of liquid hydrocarbons has been studied on Ru [1–11, 29–33], Fe [1–11, 34–40], and Co [1–11, 24, 41, 42] with most of the kinetics studies concerning the latter two catalysts. The model mentioned in the previous section, developed primarily for methanation, also applies to the production of liquid hydrocarbons.

Recent studies indicate that the mechanism of methanation and of the synthesis of higher hydrocarbons starts with the dissociation of carbon monoxide to a surface carbide [43, 44]. This is the rate-determining step and is rapidly followed by the hydrogenation of the surface carbide to methylene which can undergo either polymerization or further hydrogenation yielding higher hydrocarbons or methane. Olefins form either by the direct desorption of unsaturated species or via the dehydration of alcohols.

Alcohols form by a modification of the surface carbide mechanism which involves the insertion of a CO group and the subsequent partial hydrogenation of the resulting acyl radical. The formation of primary alcohols suggests that desorption follows this hydrogenation rapidly without intervening rearrangement.

The reaction sequences for the hydrocarbon and alcohol syntheses are shown by steps (7-21)–(7-26) and (7–27), (7-28), respectively:

Hydrocarbon Synthesis

CO Dissociation

$$\underset{\text{M}}{\overset{\text{C=O}}{\|}} + \text{M} \longrightarrow \underset{\text{M}}{\overset{\text{C}}{\|}} + \underset{\text{M}}{\overset{\text{O}}{\|}} \tag{7-21}$$

Hydrogenation

$$\underset{\text{M}}{\overset{\text{O}}{\|}} + 2\,\text{H} \longrightarrow \text{H}_2\text{O} + \text{M} \tag{7-22}$$

$$\underset{\text{M}}{\overset{\text{C}}{\|}} + 2\,\text{H} \longrightarrow \underset{\text{M}}{\overset{\text{CH}_2}{\|}} \tag{7-23}$$

Polymerization

$$\underset{\text{M}}{\overset{\text{CH}_2}{\|}} + n\underset{\text{M}}{\overset{\text{CH}_2}{\|}} + \text{H} \longrightarrow \underset{\text{M}}{\overset{\text{R—CH}_2}{|}} + n\text{M} \tag{7-24}$$

Hydrogenative desorption

$$\underset{\text{M}}{\overset{\text{CH}_2}{\|}} + 2\,\text{H} \longrightarrow \text{CH}_4 + \text{M} \tag{7-25}$$

$$\underset{\text{M}}{\overset{\text{R—CH}_2}{|}} + \text{H} \longrightarrow \text{RCH}_3 + \text{M} \tag{7-26}$$

Alcohol Synthesis

$$\underset{\mathrm{M}}{\underset{|}{\mathrm{R{-}CH_2}}} + \underset{\mathrm{M}}{\underset{|}{\mathrm{C{=}O}}} \longrightarrow \mathrm{R{-}CH_2{-}}\underset{\mathrm{M}}{\underset{|}{\mathrm{C{=}O}}} + \mathrm{M} \tag{7-27}$$

$$\mathrm{R{-}CH_2{-}}\underset{\mathrm{M}}{\underset{|}{\mathrm{C{=}O}}} + 3\,\mathrm{H} \longrightarrow \mathrm{RCH_2CH_2OH} + \mathrm{M} \tag{7-28}$$

A variation of the CO dissociation mechanism is that advocated earlier by Eidus [4], which starts with the reduction of chemisorbed CO (M=CO) to $M{=}CH_2$, via the surface enol species M=CHOH [45, 46], and is followed by polymerization and desorption according to (7-24) and (7-26).

An alternate mechanism postulates [24, 41, 42] that the CO is first reduced to the intermediate $M{-}CH_2OH$ shown in reaction (7-15), which then polymerizes to higher hydrocarbons by the elimination of water.

A final point worth noting is that Fe catalysts give much higher yields of CO_2 than Co, Ru, or Pt catalysts. For the latter, most of the oxygenated by-product tends to be water. For economic reasons, water formation is preferable because its removal from the product stream is less costly than the removal of CO_2. According to Dry and co-workers [40], the formation of CO_2 over Fe catalysts is caused by the high activity of these catalysts for the water–gas shift reaction.

7.1.4 Implications of Kinetic Studies

On the basis of the mechanism established for CO/H_2 synthesis one can assess the utility of various innovations in catalysis solving the key problem synthesis liquid hydrocarbons: selectivity control.

Clearly, the bond strength between CO and H_2 and the surface of the catalyst significantly affect catalytic activity for the simple methanation reaction [8, 9]. This is caused by two effects: alteration of the stability of the surface complex, and changing surface concentrations of CO and H_2. Both of these phenomena should also affect product selectivity for the synthesis of liquid hydrocarbons. Indeed, there is evidence that this occurs with strongly basic promoters, such as K_2O, extensively used to modify the activity and selectivity of synthesis catalysts [34–40]. These promoters are thought to donate electrons to the surface metal atoms thereby strengthening the C–metal bond of the adsorbed CO since CO acts as an electron acceptor. This modification of the catalyst can in turn affect the reactivity of the surface complex formed on hydrogenation. The selectivity of Fe catalysts is shifted toward the production of higher molecular weight liquids by promotion with K_2O because the increased Fe–C bond strength increases CO surface coverage as well as the probability of chain growth. In contrast to CO, hydrogen behaves as an electron donor on iron surfaces, and therefore K_2O promotion results in decreased hydrogen adsorption [38]. This reduces the hydrogenation activity of the catalyst and favors olefin formation. The latter probably takes place by dehydration of an adsorbed oxygenated intermediate. These concepts are consistent with observed changes in heats of adsorption of CO and H_2 on K_2O-promoted iron catalysts. Promotion increases the adsorption energy of CO and decreases that of H_2 [38, 40]. In some instances, promotion by K_2O can increase the binding energy of CO to such an extent that carbon formation occurs [38].

Accordingly, it is useful to consider the means available for changing the binding energy of CO and H_2 which control hydrogenation activity and thereby chain length. They include promotion by alkali and alkaline earths and the use of multimetallic catalysts, metal–support interactions, and crystallite size effects. These concepts already discussed in Chapters 2 and 3 will be applied in more detail to CO/H_2 synthesis in Part III.

If the active intermediates in chain growth are multiply bonded to the catalyst surface, bimetallic alloys and cluster catalysts could affect chain length since they contain the active metal constituent diluted by an inert metal matrix. Interruption of a uniform metal surface by decreasing particle size or by selective poisoning would lead to a similar effect. However, if chain growth occurs on a single site, these effects should be minimal.

7.2 WATER–GAS SHIFT REACTION

7.2.1 Kinetics

In connection with coal conversion, the primary interest in the water–gas shift reaction

$$CO + H_2O \rightleftharpoons CO_2 + H_2 \tag{7-29}$$

has centered around achieving the proper CO/H_2 ratio in synthesis gas for methanation. Most shift processes are usually run under thermodynamic equilibrium limitations which favor CO conversion at low reaction temperatures. Indeed, the last several years have seen the evolution of low-temperature shift catalysts, usually $CuO \cdot ZnO \cdot Al_2O_3$ preparations [47], which operate at 200–250°C, 150 degrees lower than do the older Fe or ZnO-Cr_2O_3 catalysts. Upon reduction, the Cu catalysts consist of highly dispersed Cu (~40 Å) stabilized against thermal degradation by the presence of ZnO and Al_2O_3 [47]. The main problem with the Cu catalysts is their increased sensitivity to poisons such as sulfur and chlorine as compared to that of the high-temperature catalysts. Thus in this case the need is for improved poison tolerance rather than for higher activity.

Many of the kinetic studies on chromia-promoted iron oxide catalysts [48–63] suggest that the reaction is:

(1) approximately proportional to the CO concentration,
(2) retarded by CO_2,
(3) almost independent of the H_2 concentration, and
(4) independent of the H_2O concentration when in excess.

Oki and co-workers [58–62] concluded from their extensive tracer work that two related mechanisms are most plausible:

Mechanism I

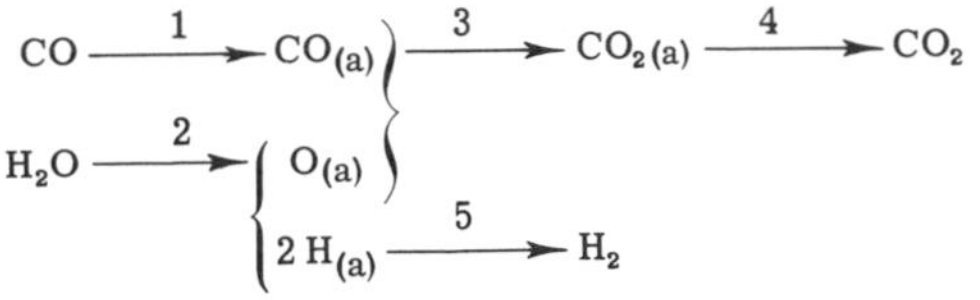

$$CO \xrightarrow{1} CO_{(a)}$$
$$H_2O \xrightarrow{2} O_{(a)} + 2\,H_{(a)}$$
$$CO_{(a)} + O_{(a)} \xrightarrow{3} CO_{2\,(a)} \xrightarrow{4} CO_2$$
$$2\,H_{(a)} \xrightarrow{5} H_2$$

Mechanism II

$$CO \xrightarrow{1} CO_{(a)}$$
$$H_2O \xrightarrow{2} OH_{(a)} + H_{(a)}$$
$$CO_{(a)} + OH_{(a)} \xrightarrow{3a} HCOO \xrightarrow{3b} CO_{2\,(a)} + H_{(a)}$$
$$CO_{2\,(a)} \xrightarrow{4} CO_2$$
$$H_{(a)} + H_{(a)} \xrightarrow{5} H_2$$

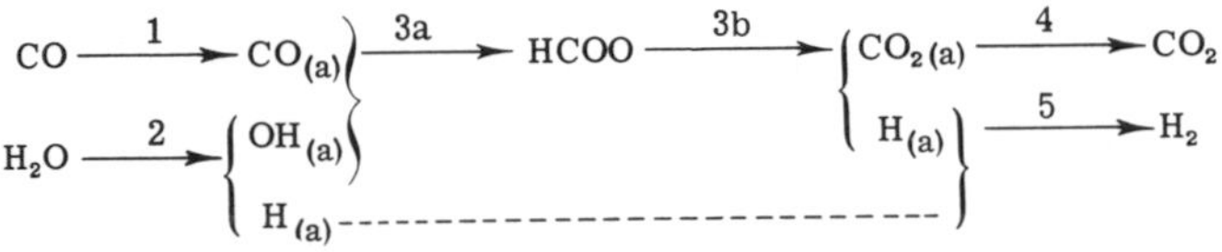

The stoichiometric number approach used by these investigators led them to postulate two rate-determining steps; namely, the adsorption of CO and the associative desorption of hydrogen. In the initial stages of reaction, CO adsorption is considered rate determining while, close to equilibrium, both steps become important since the value of the two rates (steps 1 and 5) are comparable at these conditions.

7.2.2 Technical Implications

As in the case of CO/H_2 synthesis reactions, the interaction of CO and H_2 plays a key role in determining the rate-limiting step for the water–gas shift reaction. In view of the above mechanisms, the binding energy of both reactants to the surface should be important in this respect, especially near thermodynamic equilibrium. As previously mentioned, these binding energies can be readily changed by a number of techniques, but it is also necessary to activate adsorbed water and to minimize CO_2 adsorption and H_2 dissociation. CO_2 competes with H_2 and CO for adsorption sites, acting essentially as a poison. Facile H_2 dissociation is likely to enhance the rate of approach to equilibrium for the reverse reaction. This may be a problem because, close to thermodynamic equilibrium, the rates of steps 1 and 5 are comparable, and therefore both can influence the measured rate.

The novel Cu water–gas shift catalysts developed over the past five years approach optimization of the above-mentioned parameters since they sufficiently chemisorb CO and activate H_2O, but do not dissociate H_2 readily. These catalysts are therefore more active and operate at lower temperatures, thus favoring high equilibrium conversions to CO_2 and H_2. Unfortunately, the catalysts must be meticulously guarded from poisons such as sulfur and chlorine. The former gives rise to immediate poisoning; the latter converts the stabilizing ZnO to $ZnCl_2$ and thus permits the growth of Cu crystallites. Sulfur-tolerant catalysts for low-temperature shift conversion developed recently [64–68] contain an alkali metal component such as K_2O or Cs_2O and an effective dehydrogenation component, such as Re_2O_7, Re_2S_7, or $CoMoO_4$. In the presence of water, the alkali component is believed to exist as a liquid phase in contact with the surface of the dehydrogenation function. The aqueous alkali metal phase then converts CO to a formate intermediate which is subsequently dehydrogenated to CO_2 and H_2. The interaction of these two functions apparently brings about a synergistic enhancement in catalytic activity. The primary advantage of such catalysts is that they may be used at low temperatures and at substantial sulfur levels (e.g., 1% of H_2S). They are presently effective in the 270–330°C range [64].

7.3 HYDRODENITROGENATION

Hydrodenitrogenation (HDN) is of interest and importance for petroleum operations because it is a means for removing basic heterocyclic nitrogen compounds which can poison acidic catalysts used in refining. The emerging synthetic fuels industry provides a second motivating factor for nitrogen removal: the need for minimizing NO_x formation during the combustion of coal or shale oil liquids. These fuels contain unacceptably high levels of nitrogen. So far, nitrogen removal has not received nearly as much attention as has desulfurization because sulfur, a severe catalyst poison and a serious atmospheric pollutant, has historically been the primary concern in processing petroleum feedstocks.

HDN is accomplished by the use of conventional hydrodesulfurization (HDS) technology with a slightly modified catalyst at higher temperatures and pressures and lower space velocities made necessary by the slowness of HDN [69–71].

7.3.1 Mechanism Studies

With the increased need for more effective HDN processes some work has been carried out with model compounds to gain information on key mechanistic steps, and to find some leads towards catalyst improvements.

Most of the work on monocyclic N compounds has been done with pyridine [72–86], while pyrrole has seldom been used because of its thermal instability.

According to the consensus [78, 79], the HDN of pyridine starts with its reversible hydrogenation to piperidine and is followed by the reversible hydrogenolysis of one of the C—N bonds in piperidine to yield *n*-pentylamine. The rate-determining step (RDS) of the overall HDN is the hydrogenolysis of the amine to pentane and ammonia. It has been suggested [80] that the hydrogenation of pyridine is the RDS, but most of the data obtained at conditions of facile HDN fail to support this conclusion.

According to Sonnemans and co-workers [72–74, 76, 77] the straight sequence of reactions mentioned above is complicated by the reaction of pentylamine (MPA) with piperidine (PIP) and with itself resulting in alkyl transfer and yielding *N*-pentylpiperidine and dipentylamine, respectively, and by the hydrogenolysis of these amines. Thus the following mechanism has been suggested:

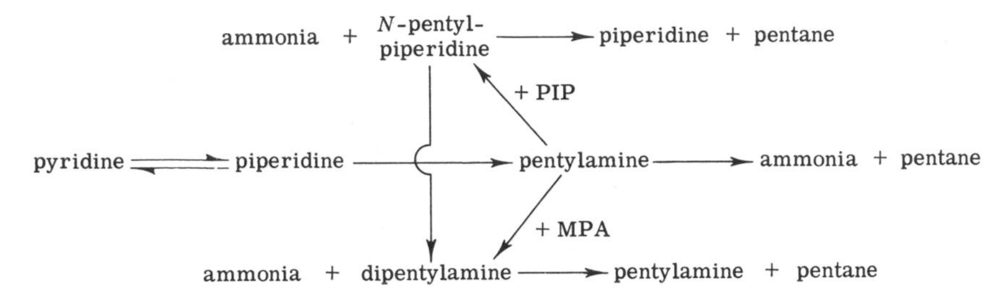

The hydrogenolysis reactions of pentylamine, dipentylamine, and pentylpiperidine yield pentane, but, in the case of the latter two amines, also products which can participate in the alkyl transfer reactions. Therefore the overall C–N hydrogenolysis, the rate-determining step, is also influenced by the equilibrium constants for the alkyl transfer reactions.

The conversion of pyridine at high pressures of hydrogen is similar in both activity and selectivity on $CoO–MoO_3/Al_2O_3$ and MoO_3/Al_2O_3 catalysts [73]. This may indicate that CoO merely adds hydrogenation activity, and that the rate-determining step, the C–N hydrogenolysis, occurs predominantly on Mo sites. If this is correct, high Co/Mo ratios are expected to give a lower HDN activity. The following surface mechanism was suggested for C–N hydrogenolysis:

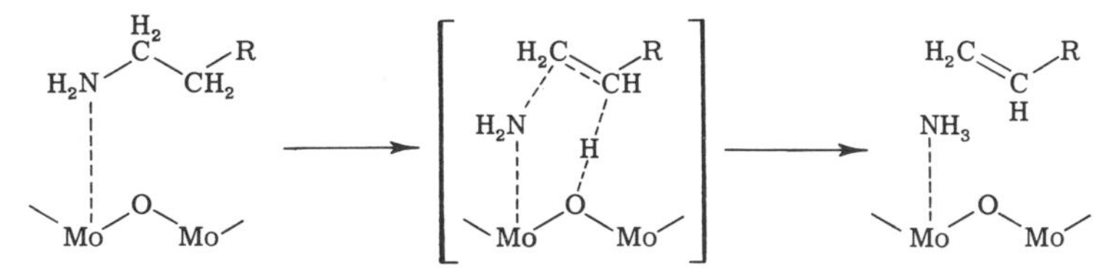

The NH_3 desorbs and the olefin is quickly hydrogenated. This sequence is supported by the fact that trace amounts of pentenes have been observed during pentylamine hydrogenolysis [76]. This mechanism suggests that increasing the Mo–N interaction should increase the rate of HDN.

In the past, kinetic studies with polycyclic nitrogen compounds have been difficult to perform because of side reactions, instability of intermediates, and analytical difficulties. Differential reactor conditions and modern analytical procedures should alleviate these problems. Only very few studies have been carried out in the last decade [75, 83] and most of the available data are from older work [80–82, 84].

Doelman and Vlugter [81] studied the hydrodenitrogenation of quinoline on unsulfided cobalt molybdate catalysts at 300–850°F and 600–1800 psi and identified tetrahydroquinolines, anilines, and aliphatic amines in the products, and concluded that these are intermediates. According to the reaction sequence postulated, quinoline is hydrogenated fast to 1,2,3,4-tetrahydroquinoline (THQ), which subsequently hydrocracks to either *o*-propylaniline or 3-phenylpropylamine. The final step was postulated to be the HDN of the aniline or amine to NH_3 and hydrocarbons. By comparing HDN rates for quinoline with those for aniline and amines, these workers concluded that the hydrogenolysis of the C—N bond in aniline is the rate-determining step in the HDN of quinoline.

Tsusima and Sudzuki [87] observed decahydroquinoline (DHQ) as well as the tetrahydroquinolines (THQ) over nickel-type catalysts. Their work and that described above suggests that the following equilibria may occur during HDN of quinoline (Q):

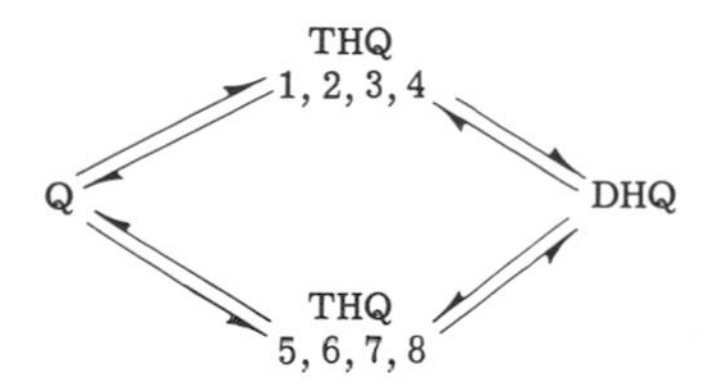

The 1,2,3,4-THQ is then hydrocracked to give alkyl-anilines which subsequently undergo HDN, the rate-determining step. DHQ hydrocracks to primary and secondary amines which can then denitrogenate to either alkylbenzenes or alkylcyclohexanes, depending on the hydrogen pressure.

In general, methyl substitution on the quinoline ring makes hydrogenation of that ring more difficult [88].

A recent study of HDN of quinoline and indole over unsulfided cobalt molybdate catalysts suggests that these heterocycles show different rate-determining steps [75]:

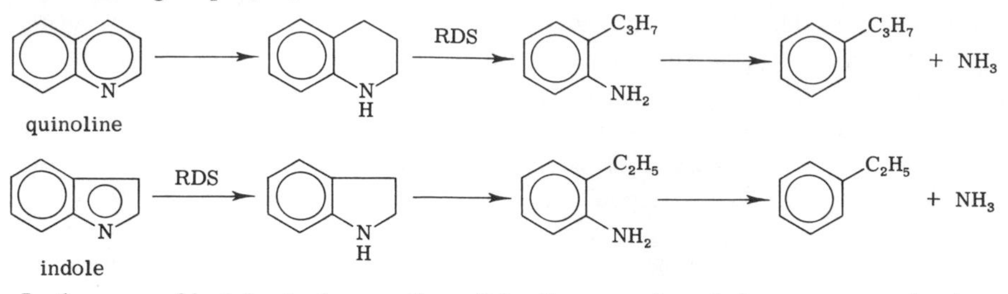

In the case of indole, hydrogenation of the five-membered ring appears to be the rate-determining step; whereas, for quinoline, hydrocracking of the hydrogenated heterocyclic ring is reported as the slow step. As the work in these systems involved no kinetic study, but was conducted at high conversion levels in an autoclave, product inhibition and other complications cannot be dismissed entirely.

It appears that there are interactions between HDS and HDN reactions [78, 79] resulting in unusual synergistic effects.

Goudriaan and co-workers [89] studied the HDN of pyridine over a cobalt molybdate catalyst at 250–400°C and 1200 psig. They found that, at high conversion levels, the temperature required to attain a certain degree of nitrogen removal is about 25°C lower with a sulfided catalyst than with the catalyst in its oxide form. The presence of H_2S results in a further reduction of this temperature requirement by 60°C. The hydrogenation of pyridine and the ring opening of piperidine occurred faster on the sulfided catalyst than on the oxide catalyst. The presence of H_2S had little effect on the hydrogenation but accelerated significantly the ring-opening reaction.

Similar effects were reported by Mayer [78] who found that nitrogen compounds severely inhibit HDS of thiophene on sulfided cobalt molybdate catalysts, but that H_2S promotes HDN activity. The explanation given for the latter effect is that H_2S increases the hydrocracking activity of the catalyst, and therefore the rate of cracking of piperidine to *n*-pentylamine, which is assumed to be the rate-determining step.

7.3.2 Implications for Further Research

While hydrodenitrogenation of cyclic compounds is not fully understood, two important findings are noteworthy. First, hydrogenolysis of C—N bonds is a key step for many HDN reactions and indeed may generally be rate determining. Second, the degree of presulfiding and the presence of H_2S can have significant promotional effects on HDN reactions. Each of these effects should be exploited to obtain more active and selective HDN catalysts.

In the preparation of improved catalysts for C–N hydrogenolysis, bifunctional systems are expected to play a significant role. Thus controlled interactions of an acid function with the basic nitrogen atom, and of an adjacent metal site with the carbon atom could lead to substantial weakening of the C—N bond in the heterocycle and perhaps to increased activity. Metal–support interactions and novel sulfides, oxides, and oxysulfides (see Part II) are likely to be important for this aspect of HDN research.

7.4 HYDRODESULFURIZATION

7.4.1 Kinetic Effects

Coal liquids contain substantial amounts of sulfur compounds which are similar to those present in petroleum crudes and include thiophenes, benzothiophenes, and naphthobenzothiophenes [90].

The literature concerning HDS chemistry and catalysis has been reviewed in some detail elsewhere [91–93]. Significant contributions to the kinetics of the HDS of monocyclic and straight-chain S compounds have been made by Kemball and co-workers [94–97] using MoS_2 and WS_2, and by Amberg *et al.* [98–103] using Cr_2O_3, MoS_2, and cobalt molybdate as catalysts. Also noteworthy is the work on polycyclic S compounds by Givens and Venuto [104].

At high partial pressures of hydrogen, the rate of HDS is approximately first order with respect to the sulfur compound. The rate and order decrease with increasing concentration of the sulfur compound, and the order can assume negative values at high S concentrations as has been observed in thiophene desulfurization over Ni_3S_2 [105, 106] or MoS_2 [105, 107]. This suggests that at high surface coverage the sulfur compound acts as an inhibitor.

The hydrogenolysis of thiophene over MoS_2 and Ni_3S_2 is slower than that of tetrahydrothiophene, indicating the possibility of two-point adsorption for thiophene and of single-point adsorption for its saturated analog [105].

The rate of desulfurization usually decreases with increasing molecular weight of the sulfur compound. In some cases this can be attributed to steric effects [104, 108], in others to electronic factors [104]. It can also be caused by catalyst deactivation, which occurs for higher molecular weight species.

The rate of desulfurization generally increases with increasing partial pressure of hydrogen [91, 98–103].

H_2S at the 2–3 mole percent level slightly decreases the rate of HDS of thiophene over MoS_2, Cr_2O_3, and cobalt molybdate catalysts, but inhibits HDS activity at higher concentrations [100, 101]. H_2S also inhibits the hydrogenation of butene, but not that of butadiene [98, 101]. Since this effect is greater than the overall decrease in the HDS rate, it appears that H_2S competes for butene hydrogenation sites [98, 101].

On the addition of oxygen to thiophene the rate of HDS over MoS_2 doubles initially but then decreases over a period of hours to the original value without any change in product composition [100]. The rate of hydrogenation of butene, an intermediate in thiophene HDS, increases one hundredfold on oxygen addition. This effect, yet unexplained, has been attributed to the formation of a more active oxysulfide surface [109].

A number of compounds reduce the rate of hydrogenolysis of thiophene over MoS_2. Their effect decreases in the order cyclopentadiene $> CS_2 > C_2H_4 > H_2S > CS_2 > CH_4 > CO$ [105] and is reversible, except in the case cyclopentadiene, which has molecular dimensions similar to those of thiophene. It has been suggested that the former adsorbs by a two-point mechanism on the same sites which activate thiophene and reduces its surface concentration.

Kinetic studies of HDS reactions over MoS_2 and WS_2 by Kemball *et al.* [95] show that the relative ease of scission of chemical bonds decreases in the series H—H > H—S > C—S > C—H > (C_2H_6) > C—H (CH_4) > C—C. The C—S scission in straight-chain compounds is easier than in cyclic compounds [97].

7.4.2 HDS of Heterocyclic Compounds

Most of the mechanistic and kinetic work in the area of HDS has been done with thiophene, the simplest model for heterocyclic S compounds present in petroleum and coal [90, 92].

Kemball and co-workers [94–97] found that the rate-determining step for desulfurization to saturated hydrocarbons is the combination of an adsorbed alkyl radical with a chemisorbed hydrogen atom to form an alkane which rapidly desorbs [97].

Amberg and co-workers [98–103] studied thiophene desulfurization over chromia and cobalt molybdate and determined the relative rates for the various steps at 415 and 400°C, respectively. On both catalysts, the initial step in HDS is the C—S bond cleavage leading to butadiene, and the rate-limiting step is the reaction of adsorbed olefin with hydrogen to form butane. H_2S inhibits the HDS of thiophene and the hydrogenation of butenes, but has very little effect on the cis–trans or double-bond isomerization, or on the hydrogenation of butadiene. These results suggest that more than one kind of sites are involved in HDS. The identification of these sites would be useful for the development of more active and selective HDS catalysts.

Among the very few kinetic studies done with polycyclic S compounds [110, 111], the investigation of Givens and Venuto is perhaps the most significant [104]. Using a cobalt molybdate catalyst in a flow system at 400°C and atmospheric pressure these authors found that benzothiophenes are much less reactive in HDS than thiophenes. Usually common reaction paths are involved which include the following reactions in the case of 3,7-dimethylbenzo[b]thiophene:

(a) alkyl group migration on the thiophene ring,
(b) dealkylation of the thiophene ring,
(c) sulfur removal from rearranged or demethylated secondary products, and
(d) direct sulfur removal from the primary reactant.

These reactions are shown in the following sequence:

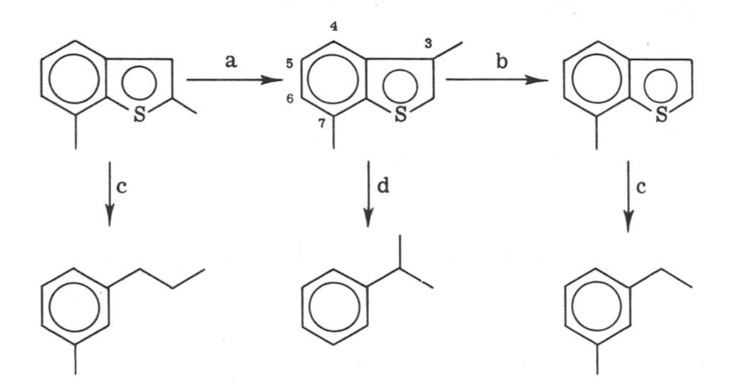

The dealkylation of, and methyl migration on, the benzene ring were relatively slow. Conversion and selectivity to various products were found to be highly dependent upon the number and position of the alkyl groups. Alkyl groups on the thiophene ring lowered conversion more than those on the benzene ring.

The formation of dihydrobenzothiophenes at high space velocities suggested that these compounds are intermediates. The initial step of the desulfurization involves the breaking of C—S bonds which can occur without the hydrogenation of the aromatic ring in the case of aromatic S bonds. No evidence was found for significant amounts of C—C bond scission.

7.4.3 Implications for Further Research

The work in the HDS area indicates that hydrogenolysis and hydrogenation reactions proceed on separate catalytic sites. Selective poisoning experiments by Amberg and co-workers suggest that hydrogenation reactions occur on strongly electrophilic sites, whereas desulfurization reactions proceed on weakly electrophilic sites [103]. This concept should be explored further and verified. If true, it suggests that more active catalysts will require a bifunctional structure and could be developed by judicious changes in catalyst–support interactions, surface acidity, and crystallite size.

The nature of surface sulfidation and its relationship to HDS catalysis is another important area that needs further study. New materials and novel catalyst preparative procedures, especially those providing higher surface areas, could play a key role in this area.

As an example, materials of the general stoichiometry $M_2Mo_3O_8$ (where M = Mg, Zn, Co, Mn, and others) have been suggested for study as HDS catalysts [109] because of the unusual position of the Mo atoms in these structures. They are located at the apices of equilateral triangles with a Mo–Mo distance 2.53 Å (shorter than in Mo metal). These materials prepared in a high surface area form [112] were indeed found to exhibit unusual catalytic properties for hydrogenation and hydrogenolysis reactions, with activities between those observed for metals and oxides [112].

Mechanism studies can be useful in discerning the role of intermediates such as olefins and diolefins in coke formation which has important consequences in catalyst deactivation and regeneration problems.

7.5 CRACKING OF POLYNUCLEAR AROMATICS

Most of the accepted models concerning the structure of coal treat it as a highly condensed polynuclear aromatic substance. The broad objective in coal liquefaction is to crack this structure selectively with a minimum consumption of hydrogen. The function of hydrogen is to "heal" the carbon atoms involved in C—C bond scission, and also to form H_2S and NH_3 in heteroatom removal. Catalyst selectivity is important for controlling hydrogen utilization and the final product distribution. Therefore, studies of the effects of catalytic properties on the selectivity patterns for polynuclear hydrocracking reactions are useful in this respect.

The hydrocracking of polynuclear aromatic hydrocarbons has been reported to proceed through a multistep mechanism involving hydrogenation, isomerization, cracking, and rehydrogenation, in that order [113–115]. This follows from studies such as those of Cawley [116] on the hydrocracking of naphthalene over MoO_3, leading to alkylbenzenes and benzene as the main products. Similarly, the primary reactions observed during the hydrocracking of tetralin, anthracene, phenanthrene, and pyrene over nickel sulfide on silica–alumina were found to be hydrogenation, isomerization, alkylation, cracking, and paring [117–120]. The latter reaction removes methyl groups from cycloparaffin rings and eliminates them as branched paraffins, mainly isobutane. Anthracene conversion to benzene was found to take place in a stepwise manner through the formation of hydroanthracenes and naphthalenes. Much of the work described in this section was done by Qader and co-workers [114, 115, 121, 122] who carried out their studies using an autoclave at 475°C and pressures of 1000–2000 psig using CoS, MoS_2, NiS, and WS_2 physically mixed with silica-alumina, the cracking component.

In hydrocracking of naphthalene [114, 115], the first step is hydrogenation to tetralin, followed by isomerization to methylindan, which is then dealkylated to indan. Alkylated benzenes form from indan or tetralin by C—C bond scission.

The first reaction in the hydrocracking of anthracene [114, 115] is the stepwise hydrogenation to di-, tetra-, and octahydroanthracenes, in turn (see Fig. 7-1). The cyclohexane rings of tetra- and octahydroanthracenes are then isomerized to five-membered rings which subsequently crack to naphthalenes. The latter are then converted to benzenes via hydrogenation, isomerization, and cracking of one of the rings. It appears from the scheme given in Fig. 7-1 that it would be advantageous to crack the center ring in the dihydroanthracene intermediate to alkyl benzenes before it consumes additional hydrogen by further hydrogenation and dealkylation reactions.

According to Sullivan and co-workers [118–120], the principal products of hydrocracking phenanthrene are tetralin and methylcyclohexane. Essentially, three types of reactions occur:

(a) Saturation and cleavage of one of the terminal rings to form a paraffin and a bicyclic such as tetralin. This reaction is generally insignificant as paraffin yields are usually small.

(b) Ring saturation and cleavage of the central ring leads to most of the methylcyclohexane and ethylcyclohexane formed. The selective saturation and cracking of the center ring desirable for minimum hydrogen consumption has not been observed.

(c) The predominant reaction is a complex sequence of hydrogenation, cracking, and alkyl transfer, yielding principally tetrahydronaphthalene.

According to Wu and Haynes [123] center-ring cracking of phenanthrene is indicated on a Cr_2O_3–Al_2O_3 catalyst with low acid strength by the presence of significant amounts of biphenyl and 2-ethylbiphenyl among the products. Therefore, the use of controlled acidity in the presence of transitional metal cations such as chromium may be an approach to enhancing the selectivity of the desired center-ring cracking reactions.

Highly condensed polynuclear aromatics tend to be refractory and difficult to study for catalytic reactions. Such is the case for pyrene. Its hydrocracking mechanism is highly complex [114] and occurs stepwise through hydrogenation, isomerization, and cracking. Isomeric hydropyrenes, naphthalenes, tetrahydronaphthalenes, and alkylated benzenes appear among the products.

For convenience, most of the above-mentioned studies have been carried out in autoclaves. This procedure has many advantages, but it can, at high conversion levels, mask specific catalytic activities by catalyst deactivation and product inhibition. Physical characterization of the catalyst before and after reaction

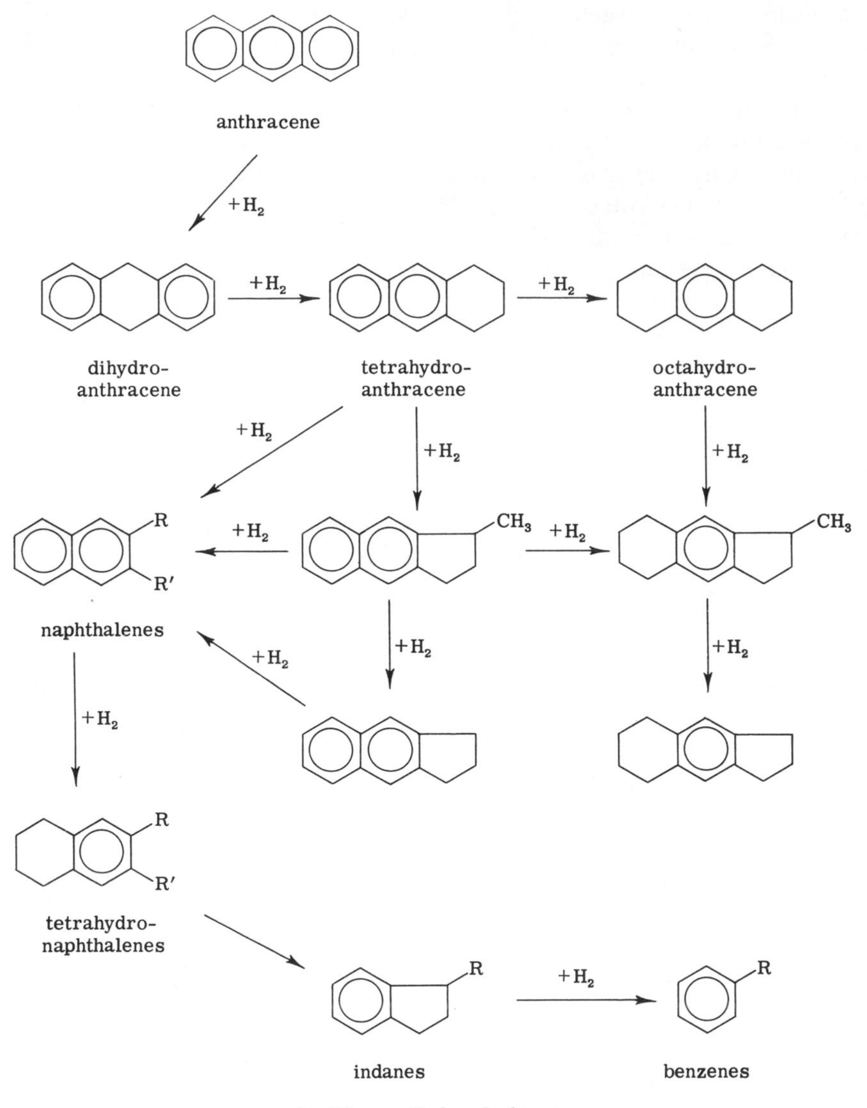

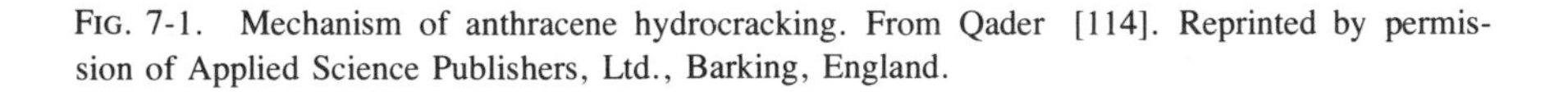

FIG. 7-1. Mechanism of anthracene hydrocracking. From Qader [114]. Reprinted by permission of Applied Science Publishers, Ltd., Barking, England.

would be useful in isolating and identifying these problems. However, this was not done in the studies by Qader and co-workers [114, 115, 122, 124] discussed above, and therefore it is difficult to evaluate relative differences between the various catalysts studied. In general, these authors found that CoS mechanically mixed with silica–alumina was the most active hydrocracking catalyst among the simple sulfides tested (MoS_2, NiS, and WS_2). This is probably due to the facile formation of saturated rings which then readily isomerize or crack. It was also found that catalysts with lower levels of alumina were more active for hydrocracking than those with higher alumina content. The lower coke yields of CoS catalysts was ascribed to their relatively higher hydrogenation activity which would prevent deposition of unsaturated coke precursors on the catalyst surface. The coke yield increased with the molecular weight of the polynuclear aromatic tested.

7.5.1 Perspective for Future Research

Studies of the hydrocracking of polynuclear aromatics give a perspective for the reactions which can occur during coal liquefaction and coal liquids refining. There is a need for similar work at differential reactor conditions on well-characterized catalysts in conjunction with mechanism studies. The ultimate objectives of this work should be to maximize the yield of lower molecular weight aromatics, and to minimize hydrogen consumption and the yield of light gases.

Optimum selectivity will certainly require multifunctional catalysis. A controlled hydrogenation and cracking activity may be central to the solution of this problem. Many of the complex oxides, sulfides, and oxysulfides discussed in Part II may well provide the desired hydrogenation activity and stability at the conditions of liquefaction or upgrading. New readily prepared and characterized solid acids [125] could give, in combination with the proper hydrogenation function, catalysts with increased activity and activity maintenance, while still retaining the desired selectivity.

7.6 CARBON AND COAL GASIFICATION

The primary objective in coal gasification is to produce high Btu gas (≥ 900 Btu/ft^3. The chemical and catalytic aspects of coal gasification cover four areas. Of these water–gas shift catalysis and methanation have already been discussed in Sections 7.1 and 7.2. The next sections will treat devolatilization, and, in greater detail, gasification.

7.6.1 Devolatilization

Devolatilization of coal is an important reaction because it forms CH_4 in the gasifier. From the thermal efficiency point of view, this is the best way to produce CH_4, in particular when compared to the gasification–methanation route. Two major reactions give rise to CH_4 formation during coal devolatilization:

$$C_nH_m \rightarrow [(4n - m)/4]C + (m/4)CH_4 \qquad \text{(hydrocracking)}$$

$$C_nH_m + [(4n - m)/2]H_2 \rightarrow nCH_4 \qquad \text{(hydrogenolysis)}$$

The first reaction can be considered to be H_2 transfer within the coal molecule. considered to be H_2 transfer within the coal molecule. The second is essentially the reaction of H_2 with C—C bonds to form CH_4.

Very little definitive catalytic work has been done on devolatilization, although it is well known that the presence of Lewis acids, such as $ZnCl_2$ or $SnCl_2$, tends to increase the yield of methane and other gases and decrease the amount of coke formed [126]. It was discovered some years ago that the gas yield depends on the temperature–time history of the coal [127]. If coal is heated at a rate of thousands of degrees per second, the yield of liquid and gaseous products can be increased in some cases to 60%. In a simplistic way, devolatilization can be considered as consisting of two parallel and competing reactions: (a) decomposition of the coal molecule to CH_4/CO/H_2, light gases, and to liquids, and (b) graphitization or polymerization reaction. The latter process is likely to have a much lower activation energy than the decomposition because it probably depends on chain steps and radical reactions. Therefore, rapid heating to high temperatures should favor the formation of light gases.

Although much of the chemistry in this area involves thermal and free-radical reactions, there may be opportunities for the use of catalysis to increase CH_4 and liquid yields, and to control the product distribution. The major problem appears to be the efficient application of a solid catalyst to a solid reactant, coal, rather than to the conventional gaseous or liquid reactant.

7.6.2 Gasification

Studies of coal or carbon gasification have mainly been concerned with the following reactions:

$$C + 2H_2 \rightleftharpoons CH_4 \tag{7-30}$$

$$2C + \tfrac{3}{2}O_2 \rightleftharpoons CO_2 + CO \tag{7-31}$$

$$C + CO_2 \rightleftharpoons 2CO \tag{7-32}$$

$$C + H_2O \rightleftharpoons CO + H_2 \tag{7-33}$$

For technological reasons, the most desirable of these is steam gasification (7-33) which, however, involves severe thermodynamic and kinetic constraints. Hydrogasification (7-30) is the other reaction of primary interest.

All four reactions mentioned have been studied in detail with and without the addition of catalysts. Much of this work is reported in the series by Walker [128]. Because of the complexity of the reactions involved and the insufficient characterization of the carbon–catalyst system, some of the mechanism studies are speculative. Dynamic changes at the coal– or carbon–catalyst interface make characterization difficult, and rate-controlling steps can vary substantially with the extent of reaction. Heat and mass transport limitation are also often encountered.

For the steam gasification of coal, the key reaction is that of steam with carbon to form CO and H_2 (7-33). This reaction is very endothermic (+31.4 kcal/mole). In most existing processes the heat for this reaction is supplied by the direct combustion of coal with air or O_2. The cost of thermal energy in this part of the process can account for as much as one third of the total process cost. The chemical problem in steam gasification is the inefficient thermal balance, as is apparent from the following reactions:

	$\Delta H_{298°K}$ (kcal/mole)	Reaction temperature
Gasification		
$2C + 2H_2O \rightarrow 2CO + 2H_2$	+62.76	1000°C (1800°F)
Shift		
$CO + H_2O \rightarrow CO_2 + H_2$	−9.83	300°C (570°F)
Methanation		
$3H_2 + CO \rightarrow CH_4 + H_2O$	−49.27	400°C (750°F)
$2C + 2H_2O \rightarrow CH_4 + CO_2$	+3.66	

Thus, the sum of the reactions which give rise to the formation of CH_4 are nearly thermoneutral. However, for reasons of thermodynamics and activity maintenance, the water–gas shift and methanation steps are carried out at much lower temperatures than is gasification. Therefore, the heat generated in these two steps cannot be efficiently utilized at the higher temperature for gasification. The solution to the problem would be catalytic gasification of coal at 800–900°F at which temperature the heat from the shift and methanation steps would be available for the gasification step.

For the steam gasification reaction alkali metal oxides and carbonates have been found to be among the best catalysts. During reaction, it is likely that the salt is in the molten state and deficient in oxygen. The molten state may lead to increased mobility and thereby enhance coal–catalyst contact for more effective gasification. According to Willson and co-workers [129], K_2CO_3 is an active catalyst for the steam reaction at 650°C and 2 atm, especially at levels of about 20%. Similar results are reported by Haynes *et al.* [130] who found that gasification rates were increased by 30–66% by using 5% alkali metal compounds.

To have a single-step gasification reactor, the gasification, water–gas shift, and methanation reactions must be catalyzed simultaneously. This would require a bifunctional catalyst. As the water–gas shift and methanation reactions are usually catalyzed by Ni or Fe, combinations of Ni and other transition metals with the alkali metal oxides or carbonates have been explored as single-step gasification catalysts [129–131]. The results are encouraging but massive amounts of catalyst are usually required for reasonable throughput. A recent patent in this area [131] describes a related process which takes advantage of the catalytic properties of alkali metal catalysts for gasification. However, instead of having a methanation constituent, increased hydrogen partial pressures are used in the gasifier. This gives rise to a substantial increase in hydrogasification (see below), and in principle methane is produced by a thermoneutral process.

Direct combination of H_2 with coal to form CH_4 occurs to a limited degree in many gasifiers, and is, because of the exothermic heat of reaction (−17.9 kcal/mole), highly desirable to reduce heat requirements for the endothermic carbon–steam reaction. Unfortunately, this reaction is very slow as shown by the relative gasification rates of carbon with H_2 and other gases, obtained by Walker and co-workers [132] at 800°C and 1 atm pressure in the absence of catalysts:

	Relative rate
$C-O_2$	10^5
$C-H_2O$	3
$C-CO_2$	1
$C-H_2$	10^{-3}

As hydrogasification is thermodynamically favored by high pressures and low temperatures, it clearly requires a catalyst. Catalytic studies [132–136] of the carbon–hydrogen reaction show some promise.

Rewick and co-workers [134] studied the change in reactivity of carbon with H_2 in the presence of various catalysts. Over an extensive temperature range, the rate of the Pt-catalyzed CH_4 formation was approximately equal to the rate of H_2 dissociation. Therefore, the catalytic effect was interpreted as an enhancement of H_2 dissociation on the metal surface, the rate-determining step [137], followed by surface diffusion across the metal–carbon interface and reaction with carbon to form CH_4.

In the range 0–1000°C and 0–1000 psig, Gardner *et al.* [133, 135] used thermobalance techniques to develop a rough kinetic model in which the activation enthalpy was assumed to be a linear function of the extent of reaction. The salts $KHCO_3$, K_3CO_3, and $ZrCl_2$ were found to be effective catalysts for hydrogasification. Another study [136] showed hydrogasification activity for Ni, Pt, and Rh. In the presence of these catalysts amorphous carbon is quickly gasified, but some of it is converted into graphite which does not gasify until higher temperatures are attained.

7.6.3 Perspective for Future Research

In steam gasification reaction, base catalysis is likely to be important for H_2O activation. Therefore correlations between basicity and gasification rates will be useful. Future programs should be directed at understanding this catalytic reaction over a broad range of temperatures and pressures using preferably high-pressure thermobalance methods. Studies of hydrogasification will have to include characterization of the catalysts and the carbons.

For the single-step gasification to methane, catalysts must be devised that combine compatible gasification and methanation functions and tolerate sulfur, nitrogen, and also mineral matter.

The advances made in catalytic materials over the last decade will be utilized in catalytic gasification. They include a better understanding of acid and base catalysis of hydrocarbon reactions, the use of metal–support interactions in the development of sulfur-tolerant hydrogenation catalysis, and applications of the refractory materials developed for catalytic oxidation or automotive catalysts.

REFERENCES

1. Pichler, H., and Schulz, H., *Chem.-Ing. Techn.* **42**, 1162 (1970).
2. Pichler, H., and Hector, A., Kirk-Othmer Encyclopedia Chem. Tech. IV, 446 (1964).
3. Nefedov, B. K., and Eidus, Y. T., *Russ. Chem. Rev.* **34**, 272 (1965).
4. Eidus, Y. T., *Russ. Chem. Rev.* **36**, 338 (1967).
5. Emmett, P. H. (ed.), "Catalysts," Vol. 4. Van Nostrand-Reinhold, Princeton, New Jersey, 1956.
6. Storch, H., Golumbic, N., and Anderson, R., "The Fischer-Tropsch and Related Syntheses." Wiley, New York, 1951.
7. Pichler, H., *Adv. Catal.* **4**, 271 (1952).
8. Vannice, M. A., *J. Catal.* **37**, 449 (1975).
9. Vannice, M. A., *J. Catal.* **37**, 462 (1975).
10. Fischer, F., Tropsch, H., and Dilthey, P., *Brennst.-Chem.* **6**, 265 (1925).
11. Mills, G. A., and Steffgen, F. W., *Catal. Rev.* **8**, 150 (1973).
12. Van Herwijnen, T., Van Duesburg, H., and DeJong, W. A., *J. Catal.* **28**, 391 (1973).
13. Lee, A. L., Feldkirchner, H. L., and Tajbl, D. G., *Am. Chem. Soc. Div. Fuel Chem. Prepr.* **14**, 126 (1970).
14. Lunde, P. J., and Kester, F. L., *J. Catal.* **30**, 423 (1973).
15. Bousquet, J. L., Gravelle, P. Ch., and Teichner, S., *Bull. Soc. Chim. Fr.* 3693 (1972).
16. Dalla Betta, R. A., Piken, A. G., and Shelef, M., *J. Catal.* **35,** 54 (1974).
17. Schoubye, P., *J. Catal.* **18**, 118 (1970).
18. Lunde, P. J., and Kester, F. L., *Ind. Eng. Chem., Proc. Res. Develop* **13**, 27 (1974).
19. Bousquet, J. L., and Teichner, S. J., *Bull. Soc. Chim. Fr.* 3687 (1972).
20. Vlasenko, V. M., and Yuzefovich, G. E., *Russ. Chem. Rev.* **38,** 728 (1969).
21. McKee, D. W., *J. Catal.* **8**, 240 (1967).
22. Baldwin, V. H., Jr., and Hudson, J. B., *J. Vac. Sci. Technol.* **8**, 49 (1971).
23. Blyholder, G., and Neff, L. D., *J. Catal.* **2**, 138 (1963).
24. Gupta, R. B., Viswanathan, B., and Sastri, M. V. C., *J. Catal.* **26**, 212 (1972).

25. Kolbel, H., Patzxchke, G., and Hammer, H., *Brennst.-Chem.* **47**, 4 (1966).
26. Subramanyam, K., and Rao, M. R. A., *J. Res. Inst. Catal. Hokkaido Univ.* **18**, 124 (1970).
27. Blyholder, G., and Neff, L. D., *J. Phys. Chem.* **66**, 1664 (1962).
28. Anderson, R. B., *in* "Catalysis" (P. H. Emmett, ed.), Vol. 4, Van Nostrand–Reinhold, Princeton, New Jersey, 1956.
29. Guyer, A., Guyer, P., Schneider, F., and Marfurt, H. R., *Helv. Chem. Acta* **38**, 798 (1955).
30. Guyer, A., Jutz, J., and Guyer, P., *Helv. Chem. Acta* **38**, 971 (1955).
31. Guyer, P., Thomas, D., and Guyer, A., *Helv. Chem. Acta* **42**, 481 (1959).
32. Karn, F. S., Schultz, J. F., and Anderson, R. B., *Ind. Eng. Chem. Prod. Res. Develop.* **4**, 265 (1965).
33. Randhava, S. S., Rehmat, A., and Camera, E. H., *Ind. Eng. Chem., Proc. Res. Develop.* **8**, 482 (1969).
34. Dry, M. E., du Plessis, J. A. K., and Leuteritz, G. M., *J. Catal.* **6**, 194 (1966).
35. Dry, M. E., and Ferreira, L. C., *J. Catal.* **7**, 352 (1967).
36. Dry, M. E., and Oosthuizen, G. J., *J. Catal.* **11**, 18 (1968).
37. Dry, M. E., Shingles, T., Boshoff, L. J., and Oosthuizen, G. J., *J. Catal.* **15**, 190 (1969).
38. Dry, M. E., Shingles, T., van H. Botha, C. S., *J. Catal.* **17**, 341 (1970).
39. Dry, M. E., Shingles, T., Boshoff, L. J., and van H. Botha, C. S., *J. Catal.* **17**, 347 (1970).
40. Dry, M. E., Shingles, T., and Boshoff, L. J., *J. Catal.* **25**, 99 (1972).
41. Sastri, M. V. C., Gupta, R. B., and Viswanathan, B., *J. Indian Chem. Soc.* **51**, 140 (1974).
42. Sastri, M. V. C., Gupta, R. B., and Viswanathan, B., *J. Catal.* **32**, 325 (1974).
43. Wentreck, P. R., Wood, B. J., and Wise, H., *J. Catal.* **43**, 363 (1976); Araki, M., and Ponec, V., *ibid.* **44,** 439 (1976).
44. Matsumoto, H., and Bennett, C. O., Div. Colloid and Surface Chem., Am. Chem. Soc.-Chem. Inst. Canada, Joint Congr., Montreal, May 29, 1977.
45. Kolbel, H., and Hanus, D., *Chem.-Ing. Tech.* **46**, 1042 (1974).
46. Schoubye, P., *J. Catal.* **14,** 238 (1969).
47. Young, P. W., and Clart, C. B., *Chem. Eng. Progr.* **69**, 69 (1973).
48. Bohlbro, H., *Acta Chem. Scand.* **16**, 431 (1962).
49. Bohlbro, H., *Acta Chem. Scand.* **17**, 1001 (1963).
50. Bohlbro, H., *J. Catal.* **3**, 207 (1964).
51. Giona, R., Passino, R., and Toselli, *L'Ingegnere* **33**, 631 (1959).
52. Hulburt, H. M., and Vasan, C. D. Sc., *AIChE J.* **7**, 143 (1961).
53. Kodama, S., Fukui, K., Tame, T., and Kinoshita, M., *Shokubai* **8**, 50 (1952).
54. Paratella, A., *Chim. Ind.* **47**, 38 (1965).
55. Shchibrya, G. G., Morozov, N. M., and Temkin, M. I., *Kinet. Katal.* **6**, 1057 (1965).
56. Podolski, W. F., and Kim, Y. G., *Ind. Eng. Chem., Proc. Res. Develop* **13**, 415 (1974).
57. Oki, S., and Mezati, R., *J. Phys. Chem.* **77**, 1601 (1973).
58. Kaneko, Y., and Oki, S., *J. Res. Inst. Catal., Hokkaido Univ.* **13**, 55 (1965).
59. Kaneko, Y., and Oki, S., *J. Res. Inst. Catal., Hokkaido Univ.* **13**, 169 (1965).
60. Kaneko, Y., and Oki, S., *J. Res. Inst. Catal., Hokkaido Univ.* **15**, 185 (1967).
61. Oki, S., Kaneko, Y., Arai, Y., and Shimada, M., *Shokubai* **11**, 184 (1969).
62. Oki, S., Happel, J., Hnatow, M. A., and Kaneko, Y., *Proc. Int. Congr. Catal., 5th* (1972) p. 173.
63. Boreskov, G. K., Yur'eva, T. M., and Sergeeva, A. S., *Kinet. Katal.* **11**, 1476 (1970).
64. Aldridge, C. L., U. S. Patent No. 3,755,556 (1973).
65. Aldridge, C. L., U. S. Patent No. 3,615,216 (1971).
66. Aldridge, C. L., U. S. Patent No. 3,539,297 (1970).
67. Aldridge, C. L., U. S. Patent No. 3,850,840 (1974).
68. Aldridge, C. L., German Patent No. 1,959,012 (1970).
69. Minachev, K. M., and Isagulyanis, G. V., *Int. Congr. Catal., 3rd, Amsterdam* (1964).

70. Smith, W. M., Landrun, T. C., and Phillips, G. E., *Ind. Eng. Chem.* **44**, 586 (1952).
71. Cottingham, P. L., White, E. R., and Frost, C. M., *Ind. Eng. Chem.* **49**, 679 (1957).
72. Sonnemans, J., Goudriaan, F., and Mars, P., *Proc. Int. Congr. Catal., 5th, 1972* **2**, 1085 (1973).
73. Sonnemans, J., and Mars, P., *J. Catal.* **31**, 209 (1973).
74. Sonnemans, J., Vandenberg, G. H., and Mars, P., *J. Catal.* **31**, 220 (1973).
75. Aboul-Gheit, A. K., and Abdou, I. K., *J. Inst. Pet. London* **59**, 188 (1973).
76. Sonnemans, J., Neyens, W. J., and Mars, P., *J. Catal.* **34**, 215 (1974).
77. Sonnemans, J., Neyens, W. J., and Mars, P., *J. Catal.* **34**, 230 (1974).
78. Mayer, J. F., PhD Thesis, M.I.T., Cambridge, Massachusetts, 1974.
79. McIlvried, H. G., *Ind. Eng. Chem., Proc. Res. Develop.* **10**, 125 (1971).
80. Flinn, F. A., Larson, O. A., and Beuther, H., *Hydrocarbon Process. Pet. Refiner* **42**, 129 (1963).
81. Doelman, J., and Vlugter, J. C., *Proc. World Pet. Congr., 6th, Frankfurt, Germany, 1963* Sect. 3, Paper 12-PD 6.
82. Schreiber, G. P., PhD Thesis, Montana State College, Boseman, Montana, 1961.
83. Koros, R. M., Bank, S., Hoffman, J. E., and Kay, M. I., *Am. Chem. Soc., Div. Pet. Chem. Prepr.* **12**, B-165 (1967).
84. Cox, K. E., and Berg, L., *Chem. Eng. Progr.* **58**, 54 (1962).
85. Wilson, W. A., Voreck, W. E., and Malo, R. V., *Ind. Eng. Chem.* **59**, 657 (1957).
86. Rosenheimer, M. O., and Kiovsky, J. R., *Am. Chem. Soc. Div. Pet. Chem. Prepr.* **12**, B-147 (1967).
87. Tsusima, S., and Sudzuki, S., *J. Chem. Soc. Jpn.* **64,** 1295 (1943); *Chem. Abstr.* 41-3801C (1947).
88. Smith, H., *in* "Catalysis" (P. H. Emmett, ed.), Vol. 4. Van Nostrand–Reinhold, Princeton, New Jersey, 1957.
89. Goudriaan, F., Gierman, H., and Flugter, J. C., *J. Inst. Pet. London* **59**, 40 (1973).
90. Akhtar, S., Sharkey, A. G., Shultz, J. L., and Yavorsky, P. M., Organic Sulfur Compounds in Coal Hydrogenation Products, presented at 167th Nat. Meeting Am. Chem. Soc., Los Angeles, California, March 31–April 5, 1974.
91. Schuit, G. C. A., and Gates, B. C., *AIChE J.* **19**, 417 (1973).
92. Schuman, S. C., and Shalit, H., *Catal. Rev.* **4**, 245 (1970).
93. McKinley, J. B., *in* "Catalysis" (P. H. Emmett, ed.), Vol. V, p. 405. Van Nostrand–Reinhold, Princeton, New Jersey, 1957.
94. Wilson, R. L., and Kemball, C., *J. Catal.* **3**, 426 (1964).
95. Kieran, P., and Kemball, C., *J. Catal.* **4**, 380 (1965).
96. Wilson, R. L., Kemball, C., and Galwey, A. K., *Trans. Faraday Soc.* **58**, 583 (1962).
97. Kieran, P., and Kemball, C., *J. Catal.* **4**, 394 (1965).
98. Owens, P. J., and Amberg, C. H., *Can. J. Chem.* **40**, 941 (1962).
99. Owens, P. J., and Amberg, C. H., *Can. J. Chem.* **40**, 947 (1962).
100. Kolboe, S., and Amberg, C. H., *Can. J. Chem.* **44**, 2623 (1966).
101. Owens, P. J., and Amberg, C. H., "Solid Surfaces and the Gas–Solid Interface," p. 182. American Chemical Society, Washington, D.C., 1961.
102. Desikan, P., and Amberg, C. H., *Can. J. Chem.* **41**, 1966 (1963).
103. Desikan, P., and Amberg, C. H., *Can. J. Chem.* **42,** 843 (1964).
104. Givens, E. N., and Venuto, P. B., Hydrogenolysis of benzo[b]-thiophenes and related intermediates over cobalt molybdena catalysts, *Am. Chem. Soc. Pet. Chem. Prepr.* **15**, A183 (1970).
105. Griffith, R. H., Marsh, J. D. F., and Newling, W. B. S., *Proc. Roy. Soc. London* **A197**, 194 (1949).
106. McKinley, J. B., *in* "Catalysis" (P. H. Emmett, ed.), Vol. 5, p. 471. Reinhold, 1957.

107. Badger, E. H. M., Griffith, R. H., and Newling, W. B. S., *Proc. Roy. Soc. London* **A197**, 184 (1949).
108. Hoog, H., *J. Inst. Pet.* **36**, 783 (1950).
109. Mitchell, P. C. H., "The Chemistry of Some Hydrodesulfurization Catalysts Containing Molybdenum." Climax Molybdenum Co., Ann Arbor, Michigan, 1967.
110. Phillipson, J. J., Kinetics of Hydrodesulfurization of Light and Middle Distillates, presented at AIChE Meeting, Houston, Texas, 1971.
111. Frye, C. G., and Mosby, J. F., *Chem. Eng. Progr.* **63**, 66 (1967).
112. Tauster, S. J., *J. Catal.* **26**, 487 (1972).
113. Brown, R. R., and Howard J., Symposium on the chemistry, occurrence, and measurement of polynuclear aromatic hydrocarbons, *Am. Chem. Soc., Div. Pet. Chem. Prepr.* **20**, 785 (1975).
114. Qader, S. A., *J. Inst. Pet.* **59**, 178 (1973).
115. Qader, S. A., and Hill, G. R., *Am. Chem. Soc. Div. Fuel Chem. Prepr.* **16**, 93 (1972).
116. Cawley, C. M., *Research (London)* **1**, 55, 33 (1947–1948).
117. Flinn, R. A., Larson, O. A., and Beuther, H., *Ind. Eng. Chem.* **52**, 153 (1960).
118. Sullivan, R. F., Egan, C. J., Langlois, G. E., and Sieg, R. P., *J. Am. Chem. Soc.* **83**, 1156 (1961).
119. Sullivan, R. F., Egan, C. J., and Langlois, G. E., *J. Catal.* **3**, 183 (1964).
120. Langlois, G. E., and Sullivan, R. F., "Refining Petroleum for Chemicals," p. 38. American Chemical Society, Washington, D.C., 1970.
121. Wiser, W. H., Singh, S., Qader, S. A., and Hill, G. R., *Ind. Eng. Chem. Prod. Res. Develop.* **9**, No. 3, 350 (1970).
122. Qader, S. A., Sridharan, R., and Hill, G. R., Abstracts of Papers, *North Am. Meeting Catal. Soc., 2nd, Houston, Texas* p. 75 (1971).
123. Wu, W., and Haynes, H. W., Jr., *Am. Chem. Soc., Div. Pet. Chem. Prepr.* **20**, 466 (1975).
124. Qader, S. A., and Hill, G. R., *Nat. Meeting Am. Chem. Soc., Div. Fuel Chem. Prepr., 160th* **14**, No. 4, Part 1, 84 (1970).
125. Boudart, M., Cusumano, J. A., and Levy, R. B., New Catalytic Materials for the Liquefaction of Coal, Rep. RP-415-1, Electric Power Res. Inst., October 30, 1975.
126. Bodily, D. M., Lee, S. H. D., and Wiser, W. H., *Am. Chem. Soc. Div. Fuel Chem. Prepr.* **19**, (1974).
127. Schroeder, W. C., U. S. Patent No, 3,030,297 (1962).
128. Walker, P. L., Jr., (ed.), "Chemistry and Physics of Carbon." Marcel Dekker, New York, 1966.
129. Willson, W. G., Sealock, L. J., Jr., Hoodmaker, F. C., Hoffman, R. W., Stinson, D. L., and Cox, J. L., "Coal Gasification," p. 203. American Chemical Society, Washington, D.C., 1974.
130. Haynes, W. P., Gasior, S. J., and Forney, A. J., "Coal Gasification," p. 179. American Chemical Society, Washington, D.C., 1974.
131. Kalina, T., and Moore, R. E., U. S. Patent No. 3,847,567 (1974).
132. Thomas, J. M., and Walker, P. L., *Carbon* **2**, 434 (1965).
133. Gardner, N., Samuels, E., and Wilks, K, *Am. Chem. Soc. Div. Fuel Chem. Prepr.* **18** (1973).
134. Rewick, R. T., Wentreck, P. R., and Wise, H., *Fuel* **53**, 274 (1974).
135. Gardner, N., Samuels, E., and Wilks, K., "Coal Gasification," p. 217. American Chemical Society, Washington, D.C., 1973.
136. Tomita, A., Sato, N., and Tamai, Y., *Carbon* **12**, 143 (1974).
137. Coulon, M., and Bonnetain, L., *J. Chim. Phys. Phys.-Chim. Biol.* **71**, 725 (1974).

PART II

ADVANCES IN SUPPORTING DISCIPLINES

Chapter 8

Reactor Engineering and Catalyst Testing

8.1 INTRODUCTION

The major developments in reactor engineering and catalyst testing fall into three areas. The first is the concept and hardware of continuous stirred tank reactors (CSTR). This type of reactor provides a method for studying the kinetics of a catalytic process at high conversion while minimizing concentration and temperature gradients. The second area is the development of a clear understanding of the effects of internal and external concentration and temperature gradients within and on the catalyst particles, respectively, on activity and selectivity. This area has been extensively treated by Satterfield and Sherwood [1] and Petersen [2], and has been reviewed more recently by Carberry [3]. Criteria have been developed to define the conditions at which mass and heat transfer limitations become important in a variety of catalytic systems. The third area covers data handling and model discrimination and has proceeded at a very rapid pace for both the mathematical treatment of complex reacting systems and the application of such techniques to specific systems. The reader is directed to reviews by Froment [4], Weekman [5], and Lapidus [6].

A great many different reactors have been developed for the study of catalytic processes. Some of these reactors operate in regimes far removed from the conditions at which the catalyst would be used in an industrial process. An example for such a reactor is the differential reactor which operates at very low conversion levels, but can supply accurate data on reactions rates, reaction order, and activation energy. Conversely, the integral reactor better simulates the real catalytic process as it operates at high conversion levels. The reactant stream composition is not constant throughout the bed of this reactor, and therefore well-defined kinetic parameters are difficult to obtain with it.

Each type of laboratory test reactor is best suited for giving certain types of information. The usefulness and applicability of test data is dependent on the skill shown in choosing a test reactor as well as in the design of test. Weekman [7], Difford and Spencer [8], and Carberry [3,9] give a detailed listing and comment on their usefulness in each stage of catalyst development.

8.2 GENERAL REVIEW OF TEST REACTORS

Catalytic reactors can be classified on the basis of a number of parameters that define the ideal limit of reactor operation important for proper measurement of reaction rates [10]. In general, industrial reactors approach these limits more or less closely. As shown in Table 8-1, the exchange of mass by the reactor with the surroundings at the limit of zero exchange defines a batch reactor, while at the other limit it specifies a flow reactor. Similarly, the total absence of the exchange of heat defines an adiabatic reactor, while complete equilibration of the reactor with the surroundings makes the reactor isothermal. Other parameters specify mechanical variables such as pressure and volume, residence time and space time behavior, all of which are important in defining the operation of a catalytic reactor.

In laboratory catalyst testing, a simplified classification system modeled on the reactors in common use is that presented in Table 8-2. The major categories are integral, differential, and pulsed microreactor where the extent of reactant conversion is used to classify reactors. Subcategories are the experimental embodi-

TABLE 8-1
LIMITING CONDITIONS OF REACTOR OPERATION[a]

Classification parameter	Limiting conditions
Exchange of mass	Batch ↔ Flow
Exchange of heat	Adiabatic ↔ Isothermal
Mechanical variables	Constant volume ↔ Constant pressure
Residence time	Unique ↔ Exponential distribution
Space–time behavior	Transient ↔ Stationary

[a]From Michel Boudart, ''Kinetics of Chemical Processes,'' © 1968, p. 17. Reprinted by permission of Prentice-Hall, Inc., Englewood Cliffs, New Jersey.

TABLE 8-2

CONDENSED SUMMARY OF CATALYST TEST REACTORS AND THEIR PRIMARY AREAS OF APPLICATION

Reactor	Catalyst selecion (screening)			Kinetic information	Reactor design	Process optimization
	Activity	Selectivity	Life			
Integral-plug flow[a]						
Adiabatic	GA	GA	GA		SI	SI
Isothermal	GA	GA	GA	GA	GA	GA
Integral-gradientless						
Well stirred continuous (CSTR)	GA	GA	GA	GA	GA	GA
Differential						
Single pass	GA	GA	GA	GA	GA	
Batch recycle	GA	GA		GA	GA	
Pulsed micro	GA	GA				

[a] GA, generally applicable; SI, applicable only in specific instances.

ments of these categories. Each of these reactor types will be discussed and rated in terms of applicability in obtaining test data for a variety of uses such as catalyst selection, kinetic information, reactor design, and process optimization.

8.2.1 Integral Reactors

The first category considered is integral plug flow reactors, which are operated at high conversion with the reactant composition changing along the catalyst bed. These reactors can be either isothermal, in which the temperature is constant and well defined throughout the catalyst bed, or adiabatic, in which the reaction heat raises or lowers the reactor temperature.

The isothermal reactor is therefore more suited for the determination of kinetic data than the adiabatic reactor. On the other hand, either type of reactor can be used in catalyst selection in which one catalyst is rated against another. The value of the comparative data or its discriminating nature will depend on the test and the parameters measured. Typical schemes of measurement are:

(a) temperature of a given conversion (e.g., 50%) with space velocity, reactor, and catalyst geometry maintained constant;

(b) light-off temperature for an exothermic reaction at which reaction rate is sufficiently high to cause the rate of heat release to exceed the rate of heat removal and the reactor temperature to jump to a high steady-state value;

(c) conversion at a given temperature with space velocity, reactor, and catalyst geometry held constant;

(d) rate parameters obtained by assuming analytical rate expression and fitting the reactor data.

Only the procedure (d) provides, on the basis of a kinetic model, data in the form of kinetic parameters related to the reactions occurring. The accuracy of the kinetic data is only as good as the assumed reaction model and simplifications. Modeling is a fine art, and good examples for its methodology can be found in the work of Froment [4] and of Lapidus [6] among others.

When using an integral plug flow reactor (see Fig. 8-1a) to screen catalysts, the best choice would be a reactor system closely resembling the actual process reactor with respect to flow properties (Reynolds number), catalyst particle size and pore size, space velocity, and reactant stream composition. Thus, for example, the best procedure for testing automotive pollution control catalysts, and one that has found widespread use, is an adiabatic integral reactor under conditions closely simulating those encountered in actual use [11].

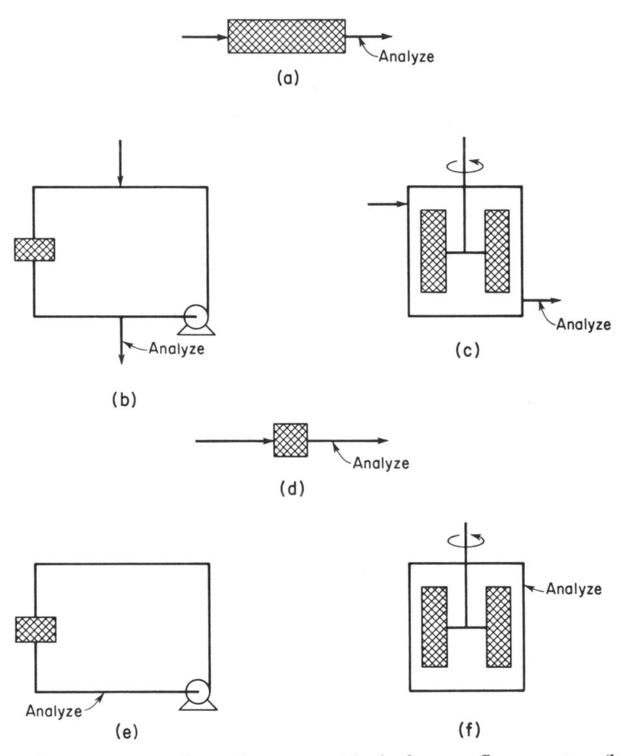

FIG. 8.1. General reactor types. Integral reactors: (a) single pass flow reactor, (b) flow recycle reactor, (c) continuous stirred tank reactor; differential reactors: (d) single pass flow reactor, (e) batch recycle reactor, and (f) stirred batch reactor.

An important point connected with integral reactors is the fact that some reactions of interest do not proceed to complete conversion because of the existence of a reverse reaction which can result in thermodynamic control.

The gradientless integral reactor is operated as a differential reactor with a low conversion per pass but at high recycling rate (see Fig. 8-1b). The low conversion per pass minimizes temperature and concentration gradients in the catalyst bed, while the high recycling rate causes considerable overall conversion.

Another gradientless reactor is the continuous stirred tank reactor (CSTR) which has been modified by Carberry [12] into a single unit with the catalyst mounted on a rotating paddle (see Fig. 8-1c). In another version developed by Berty [13], the catalyst is stationary and the gas is circulated through the catalyst bed by an impeller. To qualify as a reactor in which the gas in the system volume is perfectly mixed, a ratio of volumetric flow over the catalyst to the flow rate of reactant into the system greater than 25 is required [12]. A number of test procedures are available to determine whether sufficient mixing, and therefore gradientless operation, has been achieved [14–16].

Carberry *et al.* [15], Berty [13], Brisk *et al.* [16], and Bennett [14] discuss the detailed design, construction, and use of several types of CSTR reactors.

8.2.2 Differential Reactors

Differential reactors operate with a very low conversion of the reactants during passage through the catalyst bed. Therefore the reactant composition in the catalyst bed is assumed to be the average of the compositions of the reactants entering and leaving the reactor. Because of high flow rates, diffusion and heat transfer are also generally fast enough to prevent concentration gradients and temperature gradients in the boundary layer surrounding the catalyst particles.

The single pass differential reactor (Fig. 8-1d) is a low conversion reactor in which the reactant stream passes over the catalyst and the effluent is analyzed. Sinfelt has described the use of this reactor and its applications, especially for systems in which the catalyst activity deteriorates with time [17]. The differential batch recycle reactor (Fig. 8-1e) has a reactant stream which is contained in a loop and is continuously passed over the catalyst at a constant flow rate. The conversion per pass is differential, but the reaction products build up and the reactants are depleted with time. The recirculating stream is periodically sampled and the change of concentration with time yields the reaction rate at any time. Like the gradientless integral reactor discussed earlier, the differential batch recycle reactor can consist of separate catalyst container and pump, or can be combined in a single unit. The designs of Carberry [12] and Berty [13] are quite useful in this application as well (Fig. 8-1f).

The very small conversion occurring in the differential operation may make the analysis difficult but present-day analytical techniques and instrumentation have overcome this problem. Hydrocarbon reactions can be studied easily with capillary column gas–liquid chromatography and flame ionization detectors at extremely low concentrations. In the batch recycle reactor, each pass over the catalyst increases the total conversion in the system volume making the analysis easier. As long as the conversion per pass is small, the reactor operation remains differential. Rates at high conversion can be obtained from the batch recycle reactor, but these are transient data of conversion versus time. The CSTR is preferred because it provides stationary-state rates at high conversions and differential conditions. Versions of the Berty and Carberry reactor are produced commercially and are readily available [13].

8.2.3 Pulsed Microreactors

There is a continuous flow of carrier gas over the catalyst in the pulsed microreactor. A pulse of the reactants is injected from time to time into the carrier gas stream which then passes into a gas chromatograph for analysis. The conversion of the pulse may be small or large, but in both cases the reactant concentration over the catalyst bed is poorly defined due to mixing with the carrier gas, and reactant pulse broadening. Under special conditions, e.g., in the case of first-order kinetics, reaction rate constants can be obtained from a pulsed reaction [18]. In most other cases, adequate theoretical treatment of the system has been difficult. Thus, while pulsed microreactors are not suitable for generating kinetic parameters, they have some merit for catalyst screening because of speed and flexibility.

8.2.4 Applicability of Test Reactors

All reactors listed in Table 8-2 can be used in screening catalysts for relative activity. The measurement may give values of limited usefulness such as light-off temperature or 50% conversion temperature in the case of the adiabatic reactor, or an actual rate constant at a given temperature in the case of the differential reactors. The single pass, batch recycle, and pulsed microreactors can also provide initial rates on the catalyst in its originally prepared and characterized state before the reaction has significantly contaminated or altered its active surface. The other reactors obtain the activity measurement at some later time, and may correspond more closely to steady-state catalytic activity. An initial activity or reaction rate is desired when the main interest is the determination of reaction mechanism or reaction rates on a well-defined surface. Steady-state catalytic activity would include the phenomena of catalyst deactivation and poisoning, and will be considered under catalyst life testing. An example is catalytic hydrodesulfurization during which catalytic activity first decreases rapidly with time and then more slowly.

When a reactant can also be converted into products other than the desired one, an important parameter is selectivity defined as the ratio of desired product to total products. Selectivity measurements can be done on a comparative basis. Since process variables, such as temperature, pressure, or residence time, can affect the reaction to each product differently, the selectivity can vary with test conditions. Therefore such conditions have to be as close as possible to the industrial operating conditions in order to obtain valid selectivity data.

In the case of catalyst life testing, the necessity of maintaining conditions similar to those in industrial operations becomes even more important. Without a great deal of prior experience it would be difficult to assess the effect of unusual reactant concentrations or reactor temperatures on catalyst life. Although most process reactors do not operate differentially, the single pass differential reactor [19] and the CSTR reactor [20] can be used in some cases to follow catalyst activity during use. The batch recycle reactor and pulsed microreactor are unsuitable for testing catalyst life since their operating conditions deviate so drastically from those of industrial continuous processes.

To obtain kinetic information on a catalytic process requires knowledge of the gas composition and temperature at the catalytic surface. This effectively eliminates the adiabatic integral reactor and the pulsed microreactor in most cases. The other reactors are applicable to a varying extent, the isothermal integral reactor through a model fitting procedure and the differential reactors directly. Kinetic information is useful not only in basic studies of reaction mechanism and catalytic activity of well-characterized catalytic surfaces but also in reactor design and process optimization. Such basic kinetic data can greatly simplify the effort required in developing a process model and also reduces the amount of data needed to describe a process adequately. Other reactors can supply useful reactor design data, the only criterion being that the test in some manner simulates the actual process. In many cases the data are used in conjunction with a model of the catalytic process.

Process optimization can be considered a last fine tuning step in overall process development and reactor design, and generally requires close simulation of the industrial reactor. Tests are run on a process development unit (PDU) which is a bench scale simulation of the actual process. In this reactor the interaction of such variables as throughput, reactor length, and operating temperature with selectivity and conversion can be studied in order to maximize yield of the desired product. Of course, economic questions such as capital investment costs and feed costs enter at this point, limiting reactor size or placing a lower limit on space velocity. Smith and Carberry [21,22] describe a detailed procedure for modeling reactions and optimizing process variables to maximize the yield of product. The procedure was applied to both fixed-bed [21] and tube-wall [22] reactors for the oxidation of naphthalene, but the approach should be applicable to most reactions. Examples of the type of data obtained in this analysis include the conclusions that the fixed-bed reactor was diffusion-limited in the catalyst pellet, and that the tube-wall reactor was heat-transfer-limited at the reactor wall. Such observations suggest routes to further optimization of the process by changes in the catalyst and reactor construction, respectively.

8.2.5 Tests for Kinetic Control

The catalysts of a series subjected to comparative activity tests can differ in properties that affect the transport of reactants to, and of products from, the active surface. Such properties are catalyst particle size, pore size, thermal conductivity, and location of the catalytically active material. Even when the same catalyst base is retained to maintain particle and pore size, the distribution of the active component in the pellets may vary from catalyst to catalyst making more or less of it accessible to the reaction. It is also possible that, under the test conditions, the reaction rate is controlled by transport properties and not by catalytic activity so that the results are not just modified, but erroneous.

There are a number of theoretical and experimental test procedures for checking transport or thermal limitations in catalytic systems. Transport within catalyst particles has been treated extensively by Petersen [23], and Satterfield and Sherwood [24]. Mears [25] recently reviewed the field, discussing transport limitations in the reactor as well as in the catalyst. A common empirical test for external mass and heat transfer involves varying the flow rate at constant space velocity and noting changes in conversion. However, Chambers and Boudart [26] showed this test is not sensitive at low Reynolds numbers, and does not check for transport limitation within the particle. The latter phenomenon can be examined by the test described by Koros and Nowak [27] in which the number of catalytic sites per unit volume of catalyst is changed. This procedure can also test for transfer of heat and mass to the surface of the catalyst particle. The development of hot spots during an exothermic reaction has been treated extensively by Luss [28].

8.3 SELECTED REACTOR SYSTEMS

All of the systems discussed above concerned the reactant, generally a gas, flowing through the reactor containing the catalyst, the operation of which is well understood. Many of the reactions associated with coal conversion unfortunately do not fit into this simple category since they involve the reactions of solid or partially liquified coal with liquid and/or gaseous reactants in the presence of solid catalysts. Such multiphase systems are very complicated, and only few laboratory reactors are suitable for handling them. Their discussion follows.

Although all the basic reactor types discussed earlier are applicable in principle in multiphase reactions, significant experimental difficulties are encountered in using a number of these reactor types for liquid or solid reactants. In the case of hydrodesulfurization, preliminary studies are greatly simplified by substituting a model compound such as thiophene for coal [29,30]. However, to adequately test all aspects of catalysis during coal liquefaction and hydrodesulfurization in the definitive tests, coal or coal-derived liquids must be used.

Another simplification is possible in coal liquefaction by hydrotreating coal with a hydrogen donor solvent in one reactor and then catalytically rehydrogenating the hydrogen-depleted donor solvent in another reactor.

Because of the high pressures and temperatures involved in coal liquefaction and related processes, the most obvious laboratory procedure would be to use an autoclave as a batch reactor. However the conventional stirred or rocking autoclaves [31] have the disadvantage of requiring, because of their massive construction and size, heat-up times which are usually longer than the reaction time. This circumstance would preclude the accurate determination of reaction rates and catalyst activities.

A similar problem is encountered when the stirred tank reactors developed by Carberry [9] and Berty [13] are adapted to batch operation. In an extensive study of the solvent refining of a number of coals such a stirred autoclave was used [32], but its large thermal inertia made it difficult to heat the reactor to the reaction temperature sufficiently fast.

A better system is that described by Curran *et al.* [33] which involves the use of a small volume, low thermal inertia reactor heated in a sand bath and cooled by submersing in water. The reactor can be heated to 400°C in 2.5 min and cooled in 0.5 min and is efficiently mixed by shaking at high frequency (40 Hz). This reactor proved satisfactory in a study of the kinetics of hydrogen transfer from a hydrogen donor solvent to a coal slurry and could be used in a similar manner for the screening of catalysts for activity and selectivity. A simple measure of conversion such as hydrogen uptake would provide a measure of activity, while detailed analysis of the products would provide selectivity data.

By providing a feed system for liquid and/or slurry and an appropriate level control in the reactor, the autoclave can be operated as a flow or continuous stirred tank reactor with greatly improved capabilities. However such a system is complex, costly, and not without problems, which include storage of slurry and plugging of the feed lines and of the reactor by the slurry.

Flow reactors with a packed catalyst bed have been used to hydrogenate coal slurries with mixed results [34–36]. A number of groups are currently developing continuous stirred tank reactors for coal processing studies using the Carberry reactor as the prototype [37–39].

The data that can be obtained from such a reactor are more extensive than those derived from a batch reactor. Thus the reactor can be preheated with the catalyst, solvent, and gaseous reactant. Then the solid reactant, as a slurry, is introduced, and initial reaction rates are determined on the fresh catalyst. As the reaction proceeds, catalyst deactivation data are collected directly. At high stirring rates the reactor is well mixed, and the selectivity can be measured at several different conversion levels.

Several of the reactions important in fuel conversion, such as the synthesis of methane and hydrocarbons from CO and H_2, are highly exothermic. They require efficient heat removal to ensure good temperature control and to avoid deactivation of the catalyst and loss of product selectivity.

Many of these problems are found in the synthesis of hydrocarbons from CO and H_2. The reaction occurs rapidly at 250°C or higher, but at temperatures near 400°C the total conversion becomes equilibrium limited. Recent work has shown that a liquid flowing countercurrent to the gaseous reactants acts as a heat sink [40] and can control the temperature and permit operation at higher conversion levels and greater throughput. The major drawback of this type of operation is the reduced mass transfer of gaseous reactants through the liquid layer to the catalyst surface which may become the rate-limiting process [41].

Countercurrent reactors are under study for application to Fischer–Tropsch and methanation [40]. Work is also in progress on modeling batch liquid phase and trickle phase reactors [42–45].

An interesting concept in reactor engineering concerns reactions occurring in a medium above its critical point. Under supercritical conditions, the phase of a material cannot be described as either a gas or a liquid since the two phases become indistinguishable. Of particular interest are the unusual solvent properties of supercritical fluids and possibly their heat transfer properties [46–52].

Supercritical fluids have demonstrated their usefulness in the field of analytical chemistry because of their solvent properties. Dense gas chromatography has been used with CO_2 and NH_3 under supercritical conditions [53,54] at 40°C and 140°C and at pressures of up to 30,000 psi to separate polymers and biomolecules of molecular weight up to 400,000 by migration [53].

In the Lewis acid-catalyzed isomerization of C_4 to C_{12} paraffins, carbon dioxide has proved to be a particularly useful solvent component under supercritical conditions because of its capacity to dissolve both aluminum bromide and hydrogen. The presence of the catalyst and H_2 in the homogeneously dissolved state strongly repressed the undesired cracking that accompanies isomerization and gave the very favorable isomerization–cracking ratio of 50:1 [55].

The solvent properties of supercritical fluids have been used in coal extraction. Bartle *et al.* [56] used toluene to extract 17% of a low rank coal at 350°C and 1450 psi for analysis. The compounds present in the extract led to the conclusion that the material was not degraded during removal from the coal and that the extraction occurred under mild conditions. In a further example of supercritical fluid extraction, Wise [57] extracted the carbonaceous material from coal with benzene at 300°C and 1750 psi.

In the laboratory study of coal gasification the progress of the reaction can be followed by the weight change in the coal sample as it is gasified. This technique is generally applicable to many solid or gas–solid reactions including decompositions, absorptions, or adsorptions. Monitoring the weight change of the solid reactant has the advantage of being a direct measure of reaction without the need for analysis of all possible products.

Otto and Shelef used an atmospheric pressure microbalance to study the catalytic gasification of several coals [58,59]. This group compared the gasification of pure graphite with coal and found little difference per unit surface area which was measured in situ before and after gasification. Prior incorporation of a catalyst into the coal permitted an assessment of the catalytic effect and its deterioration with time.

For use above 300 psi, near the conditions of industrial gasification, a number of high-pressure balances are available, including quartz spring [60], electronic [61–63], and commercial units [64].

8.4 CONCLUDING REMARKS

The testing of catalytic reactions requires a careful choice of reactor systems and test conditions, in view of the complex processes encountered in coal conversion. The presence of three- and four-phase systems introduces parameters that are not properly handled by conventional laboratory reactors. These reactors can, however, be used with model compounds, and provide relevant information for such simplified systems. On the other hand, studies of coal and coal liquids require the more complex three- or four-phase reactors. Batch autoclaves can be used for the determination of catalyst activity and selectivity in coal conversion studies if special care is taken to minimize heating and cooling time by reducing their thermal inertia. They permit accurate determination of catalyst activity and selectivity. For catalyst life and activity maintenance tests a flow system is mandatory.

REFERENCES

1. Satterfield, C. N., and Sherwood, T. K., "The Role of Diffusion in Catalysis." Addison-Wesley, Reading, Massachusetts, 1963.
2. Petersen, E. E., "Chemical Reaction Analysis." Prentice-Hall, Englewood Cliffs, New Jersey, 1965.
3. Carberry, J. J., *Catal. Rev.* **3**, 61 (1969).
4. Froment, G. F., *in* "Chemical Reaction Engineering." (K. B. Bischoff, ed.), p. 1 American Chemical Society, Washington, D.C., 1972.
5. Weekman, V. W., *in* "Chemical Reaction Engineering Reviews" (H. M. Hulburt, ed.), p. 98. American Chemistry Society, Washington, D.C., 1975.
6. Lapidus, L., *Ann. Rev. Ind. Eng. Chem.* **132** (1972); Seinfeld, J., and Lapidus, L., "Mathematical Methods in Chemical Engineering," Vol. 3, Process Modeling, Estimation and

Identification. Prentice-Hall, Englewood Cliffs, New Jersey, 1973; Rossen, R. H., and Lapidus, L., *AIChE J.* **18**, (1972).
7. Weekman, V. W., Jr., *AIChE J.* **120**, 833 (1974).
8. Difford, A. M. R., and Spencer, M. S., presented at AIChE Meeting, Pittsburgh, Pennsylvania, June 2–5, 1974.
9. Carberry, J. J., and Butt, J. B., *Catal. Rev.—Sci. Eng.* **10**, (2), 221 (1974).
10. Boudart, M., "Kinetics of Chemical Processes," p. 17f. Prentice-Hall, Englewood Cliffs, New Jersey, 1968.
11. Klimisch, R. L., Summers, J. C., and Schlatter, J. C., *in* "Catalysts for the Control of Automotive Pollutants" (J. E. McEvoy, ed.), p. 1. American Chemical Society, Washington, D.C., 1975; and Schlatter, J. C., Klimisch, R. L., and Taylor, K. C., *Science* **179**, 798 (1973).
12. Carberry, J. J., *Ind. Eng. Chem.* **56**, 39 (1964).
13. Berty, J. M., *Chem. Eng. Progr.* **70**, 78 (1974).
14. Bennett, C. O., Cutlip, M. B., and Yang, C. C., *Chem. Eng. Sci.* **27**, 2255 (1972).
15. Tajbl, D. G., Simons, J. B., and Carberry, J. J., *Ind. Eng. Chem. Fundamentals* **5**, 171 (1966).
16. Brisk, M. L., Day, R. L., Jones, M., and Warren, J. B., *Trans. Inst. Chem. Eng.* **46**, 73 (1968).
17. Autoclave Engineers, Inc., Erie, Pennsylvania.
18. Makar, K., and Merrill, R. P., *J. Catal.* **24**, 546 (1972).
19. Dalla Betta, R. A., Piken, A. G., and Shelef, M., *J. Catal.* **40**, 173 (1975).
20. Mahoney, J. A., *J. Catal.* **32**, 247 (1974).
21. Smith, T. G., and Carberry, J. J., *in* "Chemical Reaction Engineering–II" (H. M. Hulbert, ed). p. 362. American Chemical Society, Washington, D.C., 1974.
22. Smith, T. G., and Carberry, J. J., *Chem. Eng. Sci.* **30**, 221 (1975).
23. Petersen, E. E., "Chemical Reaction Analysis," pp. 76, 198. Prentice-Hall, Englewood Cliffs, New Jersey, 1965.
24. Satterfield, C. N., and Sherwood, T. K., "The Role of Diffusion in Catalysis." Addison-Wesley, Reading, Massachusetts, 1963.
25. Mears, D. E., *Ind. Eng. Chem., Process Des. Dev.* **10**, 541 (1971).
26. Chambers, R. P., and Boudart, M., *J. Catal.* **6**, 141 (1968).
27. Koros, R. M., and Nowak, E. J., *Chem. Eng. Sci.* **22**, 470 (1967).
28. Luss, D., *Chem. Eng. J.* **1**, 311 (1970).
29. Owens, P. J., and Amberg, C. H., "Solid Surfaces and the Gas-Solid Interface," p. 182. American Chemical Society, Washington, D.C., 1961.
30. Kolboe, S., and Amberg, C. H., *Can. J. Chem.* **44**, 2623 (1966).
31. Gary, J. H., Baldwin, R. M., Bao, C. Y., Kirchner, M., and Golden, J. O., Removal of Sulfur From Coal by Treatment With Hydrogen, report submitted to the Office of Coal Research, U.S. Dept. of Interior, by Colorado School of Mines, 1973.
32. Wright, C. H., Perussel, R. E., and Pastor, G. R., Research and Development Rep. No. 53, U.S. Dept. of the Interior, Office of Coal Research.
33. Curran, G. P., Struck, R. T., and Gorin, E., *Ind. Eng. Chem. Process Des. Dev.* **6**, 166 (1967).
34. Kloepper, D. L., Rogers, T. F., Wright, C. H., and Bull, W. C., Research and Development Rep. No. 9, U.S. Dept. of the Interior, 1965.
35. Liquefaction of Kaiparowits Coal, Electric Power Research Institute Rep. No. 123-2, 1974.
36. Akhtar, S., Friedman, S., and Yavorsky, M., Process for Hydrodesulfurization of Coal in a Turbulent Flow Fixed-Bed Reactor, *Nat. Meeting, 71st, AIChE, Dallas, Texas,* Feb. 1972.
37. Weller, S., and Bergantz, J. A., State Univ. of New York, Buffalo, Energy Research and Development Administration Contract No. 2013.

38. Gorin, E., Continental Oil Company, Energy Research and Development Administration Contract No. 14-32-0001-1743, 1975.
39. Brooks, J., Bertolacini, R., Gutberlet, L., and Kim, D., Catalyst Development for Coal Liquefaction, EPRI Rep. AF-190.
40. Liquid Phase Methanation, U.S. Energy Research and Development Administration, Research and Development Rep. No. 78, 1974; Blum, D. B., Sherwin, M. B., and Frank, M. E., Liquid Phase Methanation of High Concentration CO Synthesis Gas, *Am. Chem. Soc. Div. Fuel Chem. Prepr.* **19**, 44 (1974).
41. Satterfield, C. N., Pelossof, A. A., and Sherwood, T. K., *AIChE J.* **15**, 226 (1969).
42. Gates, B. C., Energy Research and Development Administration Contract No. 2028, 1975.
43. Berg, L., Montana State University, Energy Research and Development Administration Contract No. 2034, 1975.
44. Greskovich, G., Air Products and Chemicals, Inc., Energy Research and Development Contract No. 2003, 1975.
45. Crynes, B. L., Oklahoma State Univ., Energy Research and Development Administration Contract No. 2011, 1975.
46. Smith, J. M., and VanNess, H. C., "Introduction to Chemical Engineering Thermodynamics," p. 58. McGraw-Hill, New York, 1959.
47. Washburn, E. W. (ed.), "International Critical Tables," Vol. 3, pp. 248–249. McGraw-Hill, New York, 1928.
48. Weast, R. C. (ed.), "Handbook of Chemistry and Physics," Vol. 52, p. D-166. Chemical Rubber Publ. Co., Cleveland, Ohio, 1971.
49. Protopopov, V. S., Juraeva, I. V., and Antonov, A. M., *Vysokotemp. Teplofiz.* **11,** 593 (1973); *High Temp. Thermophys.* **11,** 529 (1973).
50. Kaplan, Sh. G., *Inzh.-Fiz. Zh.* **21**, 431 (1971); *J. Eng. Phys.* **21**, 1111 (1971).
51. Glushenko, L. F., Kalachev, S. I., and Gandzyuk, O. F., *Teploenergetika* **19**, 69 (1972); *Thermal Eng.* **19**, 107 (1972).
52. Budnevich, S. S., and Uskenbaev, S., *Inzh.-Fiz. Zh.* **23**, 446 (1972); *J. Eng. Phys.* **23**, 1117 (1972).
53. McLaren, L., Myers, M. N., and Giddings, J. C., *Science* **159**, 197 (1968).
54. Giddings, J. C., Myers, M. N., and King, J. W., *J. Chromatogr. Sci.* **7**, 276 (1969).
55. Kramer, G. M., and Leder, F., U.S. Patent No. 3,880,945 (April 29, 1975).
56. Bartle, K. D., Martin, T. G., and Williams, D. F., *Fuel* **54**, 226 (1975).
57. Wise, W. S. D., S. African Pat. No. 69 04,202 (Jan. 9, 1970); *Chem. Abstr.* **73**, 37231j.
58. Otto, K., and Shelef, M., *Am. Chem. Soc., Div. Ind. Eng. Chem., Symp. Cataly. Convers. Coal, Pittsburgh, Pennsylvania* April, 1975.
59. Otto, K., and Shelef, M., presented at the *Int. Congr. Cataly., 6th, London* (1976).
60. McKewan, W. M., *Trans. Am. Inst. Mech. Eng.* **224**, 387 (1962).
61. Feldkirchner, H. L., and Johnson, J. L., *Rev. Sci. Instrum.* **39**, 1227 (1968).
62. Ho Bae, J., *Rev. Sci. Instrum.* **43**, 983 (1972).
63. Williams, J. R., Simmons, E. L., and Wendlandt, W. W., *Thermochim. Acta* **5**, 101 (1972).
64. Sartorius Balances, Brinkmann Instruments, Inc., Cantiague Rd., Westbury, New York 11590.

Chapter 9

Inorganic Chemistry

Inorganic chemistry has contributed to catalysis in two ways. In the broadest sense, the synthesis and characterization of new materials has provided an enormous wealth of interesting compounds covering, in addition to binary stoichiometries, a large number of ternary and higher compounds having properties applicable to catalytic phenomena. Secondly, and more specifically, inorganic chemistry applied to solids has advanced materially the solution of stability problems of catalysts arising on exposure to high temperatures and reactive environments.

Studies in solid state and inorganic chemistry encompass virtually all the elements of the periodic system in their various valence states and in their innumerable compounds. Among the vast number of compounds, there are groups of materials displaying common properties which provide a basis for an overview of their chemistry.

The most important families of compounds are shown in Table 9-1. Of these, only a limited number have been considered for catalytic applications. Furthermore, among the groups used as catalysts there are still large numbers of compositions that have not been explored. Some of these will be highlighted in the following discussion. The properties of the various groups will be summarized, and the variation in these properties between the members of a group noted. For a more detailed discussion of the properties of a large number of

TABLE 9-1

SCOPE OF INORGANIC CHEMISTRY PERTINENT TO COAL CONVERSION CATALYSIS

Class of compounds	Occurrence	Typical stoichiometries[a]	Typical examples (structures)	General comments and areas of potential application in coal conversion
Oxides				
Simple	All metals	M_3O; MO_4	Cs_3O; Al_2O_3[b]	Thermal and chemical stability except in high H_2S concentrations for certain compositions
Complex	All metals	Extensive	$CuFeO_2$ $PtCoO_2$ $FeTiO_3$ (ilmenite) $NiCrO_3$ (corundum) $BaTiO_3$ (hexagonal) $SrTiO_3$ (perovskite) $CrVO_4$ (rutile) $NiWO_4$ (wolframite) $Eu_2(WO_4)_3$ (schoelite) $Ni_2Al_2O_4$ (spinel) Al_2TiO_5 (pseudobrookite) $Se_2Ti_2O_7$ (pyrochlore) $Fe_3Al_2(SiO_4)_3$ (garnet) $NaWO_3$ (bronze) $Mg_2Mo_3O_8$	Multiplicity of oxidation states Interesting precursors for HDS catalysts New compositions for water–gas shift, in particular compounds with metal clusters Variable acidity of interest for coal liquefaction

TABLE 9-1 *(continued)*

Class of compounds	Occurrence	Typical stoichiometries[a]	Typical examples (structures)	General comments and areas of potential application in coal conversion
Sulfides				
Simple	All metals	M_3S; MS_4	V_3S (β-tungsten) VS (nickel arsenide) V_2S_3 V_3S_4 V_5S_8 VS_4 MoS_2	Thermal and chemical stability, in particular in H_2S Interesting defect chemistry that may be important for catalytic activity Complex sulfides offer new compositions of interest, containing cations of known catalytic activity Possible HDS catalysts if synthesized in high surface area
Complex	All metals	Extensive	$BaZrS_3$ (perovskite) $FeCr_2S_4$ (sulfospinel) $ZnAl_2S_4$ (sulfospinel) $Al_{0.5}Mo_2S_4$	
Carbides				
Simple	Groups I–IV, Fe, Co, Ni	M_4C; MC	Mn_4C Fe_3C (cementite) W_2C WC_{1-X}	Interstitial compounds, refractory, metallic, and hard High thermal stability
Complex	Metals	Extensive	Mo_3Al_2C (β-manganese) Pd_3AlC (perovskite) Pt_3SnC (perovskite) Pt_3PbC (perovskite) $W_{16}Ni_3C_6$ (K-carbide) V_2AlC (H-phase)	Unstable in high H_2S concentrations, but certain compositions likely to be stable in low H_2S concentrations Potential for methanation, where carbiding is a problem Complex carbides likely to retain metal characteristics, but with increased thermal stability

Nitrides				
Simple	Same as carbides	Same as carbides	Ti_2N; V_2N TiN; VN Ta_3N_5 Co_3N_2	Interstitial, similar to carbides High thermal stability, refractory and hard Unstable in high H_2S concentrations More stable than carbides in low H_2S concentrations Have been explored for Fischer–Tropsch, however no clear evidence of unusual behavior
Complex	Metals	Extensive	V_3Zn_2N (β-manganese) Ni_3AlN (perovskite) Ti_2AlN (H-phase)	Less refractory structures of interest for HDN
Borides, phosphides, silicides	All metals	M_3B; MB_{12}	Ni_3B (isolated B atoms) MoB (B chains) VB_2 (B layers) CrB_4 (3-dimensional boron frameworks) $Co_{21}Hf_2B_6$ (perovskites)	Like carbides and nitrides, high thermal stability Thermodynamic evidence for sulfur resistance in high H_2S concentrations, in particular Group VIII
		M_2P; MP_4	Co_2P (isolated P atoms) VP FeP_2 (P_2 groups) CdP_4 (phosphorus layers)	Interesting for methanation Possibly resist carbiding, in particular Group VIII Will not resist oxidizing environment at high T
		M_3Si; MSi_3	Mo_3Si (isolated Si atoms, βW struct.) TiSi (chains) $TiSi_2$ (layers) $SrSi_2$ (3-dimensional networks)	

TABLE 9-1 *(continued)*

Class of compounds	Occurrence	Typical stoichiometries[a]	Typical examples (structures)	General comments and areas of potential application in coal conversion
Alloys and multi-metallics	All metals	Extensive	Cu_3Pt (ordered only with Ni_3FC long annealing times) FeSW (nickel arsenide structure) Cu_3Al (Hume-Rothery electr. CoAl compound, Engel- CuZrW Brewer correlations) $ZrPt_3$	Ability to obtain gradual change in properties or dramatic changes in selectivity patterns Certain intermetallic compounds likely to exhibit high thermal stability All metals unstable in high H_2S Possible H_2S resistance of alloys at low H_2S Definite potential for improved methanation
Organo-metallic	—	—	$Ni_4 [CNC(CH_3)_3]_7$ (Ni clusters) $Rh_4(CO)_2$ (Rh clusters) $Pt_3(CN)_6(C_4H_9)_6$ (Pt–Pt bonds)	Well-characterized reactive centers, high activity and selectivity Anchoring on high surface area supports possible Unusual and controllable metal–metal bonds Maximum availability of active center, however, poor thermal stability Some indications of sulfur tolerance, but unconfirmed Precursors for synthesis of supported catalysts Clusters show interesting activity patterns Possibility of CO/H_2 synthesis by clusters an important aspect of the application to coal conversion

[a] Emphasis on metals (M).
[b] These are just two examples of a large variety of structures and stoichiometries.

catalytically interesting materials, the reader is referred to an earlier study [1]. Multimetallic alloys and clusters discussed in Chapter 2 will be mentioned only briefly.

9.1 OXIDES

Metal oxides constitute a very large class of compounds which include important industrial catalysts. A few representative examples are shown in Table 9-2.

Oxides exhibit a large number of structures and compositions. The examples shown in Table 9-1 are only representative of an extensive and complex chemistry.

With the exception of the volatile oxides of Os and Ru, the oxides are stable in the temperature range and environments of interest to catalysis. Under most conditions, oxides maintain high surface areas up to temperatures approaching 1000°C. For this reason such oxides as Al_2O_3 and SiO_2 are routinely used as catalyst supports. Certain environments, however, are damaging to the stability of oxides. Steam, in particular, leads to crystal growth and agglomeration at elevated temperatures [2]. This puts a constraint on the use of steam in certain processes at elevated temperatures.

The chemical stability of oxides varies across the periodic table. This is particularly evident when one examines the behavior of various oxides in an H_2S environment, especially at elevated temperatures [1]. Three degrees of stability can be distinguished. The most stable oxides include those of Al, Si, and the alkaline earth metals, Be and Mg [1]. The oxides of the rare-earth elements as well as those of Group IV (Ti, Zr, Hf) show intermediate stability. All other transition metal oxides are thermodynamically unstable at 1% H_2S, the level expected in coal liquefaction. The degree of stability depends on the temperature

TABLE 9-2
REPRESENTATIVE REACTIONS CATALYZED BY OXIDES

Reaction	Catalyst
Partial oxidation	$CoO-MoO_3$
	V_2O_5
Oxidative dehydrogenation	$Bi_2O_3-MoO_3$
Cracking	Zeolites
Reforming	Al_2O_3 (acid function)
Polymerization	MoO_3/Al_2O_3
Water–gas shift	$CuO + ZnO-Al_2O_3$
	$Fe_3O_4-Cr_2O_3$
Methanol synthesis	$CuO-ZnO-Al_2O_3$

and the partial pressures of hydrogen and hydrogen sulfide, but the relative trend remains the same for lower concentrations. It is not surprising, therefore, that oxide catalysts are susceptible to H_2S poisoning, as observed in methanol synthesis and the water–gas shift reaction.

Similar conclusions can be drawn regarding the stability of oxides in a reducing environment. The reducibility of different transition metal oxides increases as one progresses from the left to the right of the periodic table [1]. This trend persists through the Group VIII metals. Thus, the reducibility trend in the first series is Fe $<<$ Co $<$ Ni [3].

In the catalytic applications of oxides, properties of particular interest are surface acidity, which can be varied over several orders of magnitude [4–7], and their ability to exchange lattice oxygen with gaseous O_2 [8]. The former is used in the petroleum industry extensively for cracking and reforming [1], while the latter is the basis for the ability of certain oxides to catalyze partial oxidation of hydrocarbons [9].

As shown by the examples in Table 9-1, complex oxides offer a wealth of compositions and structures, many of which have received only limited attention in catalysis. The variety in the structure of complex oxides is illustrated by the difference between two groups of materials of catalytic interest: spinels and perovskites. In spinels, the basic structural unit is a close-packed network of oxygen atoms. The metal ions reside in the tetrahedral and octahedral interstices formed by this network. The type of site occupied depends essentially on the size of the cation [1]. In general, spinels are thermally very stable and exhibit a range of compositions that depend on the extent of the occupation of interstitial sites. Typical examples are $CrAl_2O_4$ (ruby) and $MgAl_2O_4$ (spinel).

In perovskites, on the other hand, one of the metal cations of the complex oxide is too large to fit in the interstitial sites. This metal becomes part of the close-packed network, with a metal–oxygen ratio of 1:3, while the other smaller cations again reside in octahedral oxygen interstices. The interesting feature of this structure is its ability to exist with less than stoichiometric amounts of the large cation, thus forming a rather open structural network. One extreme case is WO_3, where all the sites available for large cations such as Li, Na, K, and Rb are empty. The structure is nevertheless stable, and even admits smaller ions such as H^+. This ability to exist in a number of stoichiometries allows for practically continuous variations in oxidation states, and therefore continuous changes in the electronic properties of the solid.

The large number of compositions and structures result in complex oxides with a variety of electronic and chemical properties. These are likely to be reflected in changes in surface properties which may prove to be of catalytic interest. In the perovskite structure, cations of transition metals are located in unique and unusual environments. Certain oxides, for example, exhibit smaller metal–metal distances than encountered in the element itself. This is the case in $BaRuO_3$ with a Ru–Ru distance of 2.55 Å [10] or in the Mo clusters of $Mg_2Mo_3O_8$ [11]. The catalytic properties of such interesting compositions have not been studied in detail, but have been applied in certain cases to solve specific problems. The $BaRuO_3$ system provides an example of such an application. It has been known for some time that ruthenium is one of the most effective NO_x reduction catalysts, in particular because of minimum NH_3 formation [12]. It cannot be used as an automotive exhaust control catalyst, however, because it encounters oxidative conditions which volatilize the ruthenium as RuO_4. To stabilize the ruthenium during oxidation, Shelef and co-workers prepared the metal as a complex oxide with alkaline and rare-earth elements [13]. Under oxidizing conditions, the elements form a nonvolatile ruthenate. Under reducing conditions, the stable state is a two-phase system consisting of an oxide and regenerated catalytically active Ru. The solid-state chemistry of the ruthenium–alkaline earth system, therefore, lends itself ideally to the solution of the stability problem of ruthenium in an oxidizing environment.

9.2 ZEOLITES

Zeolites have been used in many catalytic reactions, and their synthesis and solid-state chemistry has been studied extensively over the last decade. They consist of SiO_4 units partially substituted by Al ions with positive ions such as Na^+ or K^+ to compensate for the charge imbalance between Al^{3+} and Si^{4+}. The SiO_4 units are combined in the silicates by sharing faces, edges, or corners. In doing so, they form open structures with cavities and channels of varying sizes. This unusual structure is largely responsible for many of the interesting properties of these materials.

Fibrous structures as eddingtonite ($Ba(Al_2Si_3O_{10})\cdot 4H_2O$), lamellar structure as phillipsite ($(K, Na)_5Si_{11}Al_5O_{32}\cdot 10H_2O$), or interconnected polyhedral cavities as chabazite ($Ca(Al_2Si_4)O_{12}\cdot 6H_2O$) are seen in naturally occurring zeolites [14]. The zeolites of the latter type have found greatest application in catalysis and can be synthesized with controlled cavity and channel sizes, thereby providing a considerable variation in physical properties. But it is the chemistry of these materials, and in particular their acidic properties, which are largely responsible for their extensive use in catalysis since their introduction as fluid-cracking catalysts in 1962 [15].

The most interesting aspect of this inorganic chemistry is the synthesis of zeolites since it determines their structure and size characteristics. In this respect, the type of hydrated cations present in the starting gels is important [16]. In general, the more open zeolite structures are prepared from gels containing sodium rather than potassium. This is a direct function of the size of the hydrated ion, since the tetrahedral building blocks of the zeolite are believed to surround this hydrated ion during synthesis [16].

Another important aspect is the stability of the zeolites which is a function of composition. For example, zeolites which substitute germanium for silicon ions are considerably less stable [17], while Si–Al zeolites deficient in aluminum tolerate severe hydrothermal conditions [18,19]. Similarly, the greater Si–Al ratio in Y zeolites compared to that in X zeolites improves structural stability.

The acidity of zeolites, a property of great interest to catalysis, is primarily controlled by the types of ions present in the structure which compensate the imbalance between Al^{3+} and Si^{4+}. Zeolites with univalent cations (e.g., Na^+, K^+) have low acidity. Ion exchange of these materials with Ca^{+2}, Mg^{+2}, Sr^{+2}, and Zn^{+2} leads to an increase in the acidity, with a distribution concentrated towards strong acid sites [20,21].

Zeolites have shown evidence for much greater poison resistance to H_2S and NH_3 than conventional amorphous silica-alumina cracking catalysts [1], the reason for which is not well understood. This resistance has important implications in coal conversion where both H_2S and NH_3 are present and can, by adsorption, affect the acidity and selectivity of the catalyst.

An aspect of zeolites which has been recognized only recently and may have interesting catalytic consequences is the existence of metal–zeolite interactions. First observed for Pt exchanged into a Y zeolite [22], and since confirmed for Rh and Pd [23], the small metal clusters in the zeolite cages are found to be electron deficient. This is concluded from ESR spectra [23] as well as the catalytic behavior of the system [22,24] which, in the case of platinum [22], resembles that typical of iridium, its neighbor in the periodic table. This ability to modify activity by varying the electronic interaction with the zeolite support has implications for the sulfur resistance of these catalysts according to Dalla Betta and Boudart [22].

9.3 SULFIDES

Formation of transition metal sulfides is particularly important in the synthetic fuels area. Many of the reactions are carried out in high concentrations of H_2S, and therefore require catalysts that withstand such environments. This is particularly true for hydrodesulfurization. In other reactions, the problem of sulfur tolerance is also severe. It is likely that many catalysts are in the sulfided or partially sulfided state during reaction. The study of sulfides is therefore an important aspect of a general review of materials.

In spite of their importance, only few mixed sulfides of Ni, Co, W, and Mo have been tested as catalysts, primarily in hydrogenation reactions in the presence of sulfur compounds [25]. Only limited information is available on the structure of these materials at reaction conditions. A detailed study of the Ni–W-sulfide system by Voorhoeve *et al*. [26] has shown that the interaction between the nickel and the WS_2 matrix has an important effect on the catalytic activity. This suggests that complex sulfides should be explored more extensively for catalytic applications, in particular in view of the growing interest in the solid-state chemistry of these materials [27–29].

The examples in Table 9-1 illustrate the large number of stoichiometries possible for vanadium sulfide. Many other metal sulfides show similar diversity. Metal sulfides often exhibit extensive homogeneity ranges (e.g., VS → V_5S_8), and the thermal stability of a particular composition depends on sulfur pressure.

The thermal chemistry of a number of metal sulfides was studied as early as the 1930s. For vanadium, for example, Biltz and co-workers investigated a broad range of compositions at various sulfur pressures [30]. These studies were followed by those of Tudo and Tridot [31,32] on the decomposition of vanadium sulfides. Besides these investigations, there has not been much work in this area, especially at the low sulfur concentrations of interest in most catalytic systems.

The general evidence is that the lower sulfides such as CoS and VS are stable at high temperatures; VS has been reported at temperatures as high as 1500°C in vacuum [33]. By contrast, VS_4 decomposes at temperatures as low as 300°C even in the presence of high sulfur pressures [31,32]. This low stability explains the decomposition of VS_4 catalysts tested for hydrogenation and desulfurization [34].

The stability of sulfides in H_2S is a function of pressure and temperature [1]. The important parameter is the equivalent (or virtual) sulfur pressure which is in equilibrium with an H_2S/H_2 mixture and determines the possibility of sulfide formation for other materials. For metals the minimum H_2S concentration for bulk sulfide formation can be calculated from free energy data [1]. The few examples given in Table 9-3 show that the sulfiding tendency of transition metals varies widely. In general, all transition metals will sulfide in the high H_2S

TABLE 9-3
MINIMUM CONCENTRATION OF H_2S (AS PERCENT H_2S IN H_2) REQUIRED FOR BULK SULFIDATION; T = 700 K[a]

Stoichiometry	Concentration (%)	Stoichiometry	Concentration (%)
CrS	4×10^{-4} (4 ppm)	RuS	1×10^{-6} (10 ppb)
MnS	2×10^{-11}	OsS	2×10^{-2}
FeS	7×10^{-2} (700 ppm)	PtS	1.3
CoS	3×10^{-2}	CuS	73
NiS	2×10^{-1}		

[a]From Boudart *et al.* [1], courtesy of the Electric Power Research Institute.

concentrations of coal liquefaction, and the reaction therefore proceeds in the presence of a metal sulfide. At lower H_2S concentrations, however, this will not be the case. For example, at methanation conditions (H_2S < 1 ppm) sulfide catalysts such as NiS are unstable, but there is still considerable interaction of H_2S with the catalyst surface.

The above observations apply to an H_2S/H_2 environment. Several sulfides (such as MnS or VS) are stable at very low H_2S pressures or even in vacuum. However, most sulfides will oxidize in the presence of O_2 [1]. This is an important aspect of the chemical stability of these materials because many catalytic processes use oxidative procedures for catalyst regeneration. In such circumstances the balance between the oxide and sulfide becomes important and the possibility exists of partial sulfidation—namely the formation of oxysulfides. Because of the lack of information on these important materials, this area requires more research. The oxidation of both bulk and supported sulfides needs study because the chemical behavior of supported systems may be different from that of the bulk, as has been suggested for cobalt molybdate HDS catalysts [35].

There are a number of complex sulfides that result from the versatile chemistry of transition metals. The layered structure of many disulfides leads to complex sulfides that resemble intercalation compounds, such as FeV_2S_4, $TiCr_2S_4$, and $CuCrS_2$ [36,37]. Compounds that have the perovskite structure MM^1S_3 (M = Sr; Ca, Pb; M^1 = Zr, Ti) have also been synthesized and studied [38]. Complex sulfides that incorporate a rare-earth metal (such as La_4NiS_7 and $La_2Fe_2S_5$) are another group of sulfides which has received recent attention [39], as have the thiospinels [40,41] which occur as nonstoichiometric compounds. $Al_xMo_2S_4$, for example, has been prepared with values of x as low as 0.5. The latter stoichiometry exhibits unusual molybdenum clusters similar to some of the complex oxides discussed in Section 9.1. Many of these recently synthesized and characterized complex sulfides contain cations as Mo^{4+} and Ni^{2+} known to be catalytically active in a sulfide matrix. They are therefore catalyst candidates for the conversion of coal in strong sulfiding environments.

9.4 CARBIDES AND NITRIDES

Carbides and nitrides of transition metals have related structures and show similarities in many of their chemical and physical properties [42]. Because of the small size of C and N compared to that of the transition metals, the nonmetal occupies interstitial sites formed by close packing of transition metal atoms and substantially modifies the properties of the parent transition metal. Some of these changes are discussed in this section.

Carbides and nitrides are known as refractory compounds. Thus, several carbides and nitrides melt or decompose above 3000°C, and TaC has the highest melting point known for any material (about 3983°C) [43]. This stability of bulk carbides and nitrides is likely to persist also in a finely dispersed form, since carbides are known to resist sintering. In fact, metallurgists have had to add binders such as cobalt to achieve dense cemented carbides for cutting tools [43, p. 9]. Carbides and nitrides are therefore a promising class of materials for potential application in such exothermic reactions as methanation and Fischer–Tropsch synthesis where sintering is a severe problem.

The resistance of WC to attack by severe environments is well known, and is one of the reasons for evaluating this compound as a battery electrode [44].

In the presence of O_2, however, the thermodynamic properties of these compounds indicate that carbides and nitrides will oxidize [1]. For kinetic reasons, this oxidation does not take place at low temperature; but, at the high temperatures of regeneration, oxidation may be a problem in the use of these materials as catalysts and requires further investigation.

Thermodynamic calculations indicate that most carbides are likely to sulfide in the presence of S compounds [42]. The sulfur level at which this occurs and the rate of sulfidation have yet to be determined experimentally. In general, the same is true for nitrides. The most stable nitrides, however, show a positive free energy of sulfide formation, and are therefore expected to survive even in severe sulfiding environments [42].

It is interesting to note that a highly refractory material does not possess necessarily improved sulfur tolerance. Tungsten, for example, will react with H_2S according to the equation

$$W + 2H_2S \rightleftharpoons WS + 2H_2$$

with a free energy change of -16.7 kcal mole^{-1} at 700 K [1]. For the carbide, on the other hand, the reaction

$$WC + 2H_2S \rightleftharpoons WS_2 + CH_4$$

has a free energy change of -25 kcal mole^{-1} [1]. The main reason for the increased tendency of WC to sulfide in the presence of H_2S in this example is the formation of CH_4.

With the exception of several Group VIII noble metals (Ru, Rh, Pd, Ir, and Pt), all transition metals form a number of carbide and nitride stoichiometries [43] which follow the empirical rules postulated by Hägg in 1931. If the radius ratio of nonmetal to metal is less than 0.59, the compounds formed have a simple, interstitial structure [1]; otherwise, more complex structures form. In addition to such binary compounds, a number of multimetallic carbides and nitrides have been synthesized, among which the Nowotny octahedral phases are of particular interest [43]. They contain two or more transition metals, and display complex structures, many of them related to the oxide structures discussed earlier. Some examples are shown in Table 9-1 and include the carbide and nitride perovskites such as Pt_3ZnC [1]. These compounds are likely to show modifications of the bulk properties of the parent metals which in turn should be reflected in the surface chemistry of these materials.

Carbonitrides and oxide-carbides of many transition metals [45] offer a broad range of physical and chemical properties for catalytic applications if efforts to prepare them with high surface areas are successful [46].

9.5 BORIDES, SILICIDES, AND PHOSPHIDES

Borides, silicides, and phosphides display many similarities in structure and properties [1,5]. They are formed by most metals, including the Group VIII noble metals.

Borides and silicides display thermal stabilities that are almost as high as that of the carbides and nitrides. Thus TiB_2 melts at 2980°C. Elements that form the most stable nitrides are also expected to form phosphides with melting points above 2200°C [47].

Silicides and borides do not show such a dramatic decrease of stability across the periodic system on approaching Group VIII as do the carbides and nitrides. The stability of the binary phases even with the noble metals of Group VIII presents the possibility of modifying the catalytic properties of these metals.

A comparison of the enthalpies of formation of borides, silicides, and phosphides with those of oxides shows clearly that the former tend to oxidize in the presence of O_2. However, in practice these compounds, and in particular the silicides, have shown very high resistance to oxygen attack [48]. They are used as heating elements because of this property. According to Searcy [48], this chemical resistance is probably the result of a coating of SiO_2 which protects the compound from bulk oxidation. However, the formation of such a coating on a metal silicide catalyst would also lead to deactivation.

The most interesting chemical property of this group of materials becomes evident upon examination of the thermodynamics of their sulfidation [1], even though the thermochemistry of these compounds and especially of the borides has not been studied very extensively [48,49]. As one proceeds from left to the right on the periodic table, silicides become more resistant to sulfidation. Group VIII silicides, in fact, are probably sulfur tolerant even in high H_2S concentrations. A similar trend is observed for phosphides, and is expected for borides in view of the similarity in other properties of these three classes of compounds [1].

The most evident common feature of borides, silicides, and phosphides is their structure. The size of phosphorus and silicon is too large to form interstitial compounds of the type formed by C and N. Boron is on the borderline of the Hägg rule [35] discussed earlier for interstitial carbides and nitrides, and in general behaves more like Si and P than C or N. The structures formed are characterized by a wide range of stoichiometries and the formation of nonmetal networks at high nonmetal to metal ratios. Some examples are illustrated in Table 9-1.

Both simple compounds and complex structures belong in this class as exemplified by the borides which forms series of compounds such as MoCoB, WFeB, WCoB, and perovskite-like structures exhibiting cluster formation in the case of $(Co_{13})(Co_8)Hf_2B_6$ [50, p. 180].

Because of its size, boron can form numerous borocarbides such as Mo_2BC and ScB_2C_2 [50]. The latter stoichiometries, also formed by the 4f elements (lanthanide series), display close metal–metal distances and interesting B–C networks [50]. They are likely to have unusual chemical and physical properties.

9.6 ORGANOMETALLIC COMPOUNDS

The wide scope and versatility of organometallic chemistry is exemplified by the extensive literature on this subject [51–53]. Among the recent developments in organometallic chemistry, catalysis by polynuclear cluster complexes and the anchoring of homogenous metallorganic catalysts to organic or inorganic polymers offer some potential applications in coal conversion.

The primary interest in the synthesis of organometallic complexes with metal clusters is the possibility of new catalytic properties not exhibited by the mononuclear species. According to Norton [54], these complexes provide the possibility of bonding with several metal atoms, thus leading to selectivity control, the migration of species on the catalyst cluster surface, and the added flexibility in achieving modifications of catalytic properties due to metal–metal interactions.

The cluster compounds recently synthesized range from small clusters of Pt and Ir compounds such as $Pt_3[CN(C_4H_9)]_6$ [55] and $Ir_3(CO)_{12}$ [56] to large cluster compounds of gold, $Au_{11}[P(C_6H_5)_3]_7I_3$ [57], and rhodium, $Rh_{13}(CO)_{24}H_3$ [58]. While the properties of these cluster compounds are not completely understood, several interesting aspects have been determined. The mobility of some of the ligands is substantial. In $(\pi\text{-}C_5H_5)_2Rh_2(CO)_3$, for example, there are two types of carbonyl groups, bridging and terminal, which are so mobile that C-13 NMR cannot distinguish between them [59]. This mobility parallels the fast surface diffusion of chemisorbed species on heterogeneous materials. There are also clear indications of metal–metal interactions in these clusters [54]. A recent report on the preparation of multimetallic clusters in organometallic complexes [60] therefore opens up the possibility of applying such materials in catalysis as has been done in the case of heterogeneous multimetallic catalysts (see Chapter 2). There are preliminary indications that clusters exhibit unique catalytic chemistry. Thus

Meutterties *et al.* [61] observed cyclization of acetylene to benzene and butadiene to cyclo-octadienes at room temperature in the presence of $Ni_4[CNC(CH_3)_3]_7$. No such reaction has been observed with the monomer which does not contain the Ni cluster network.

The main limitation to the application of homogenous catalysts to coal conversion [1] is catalyst recovery. When such a precious metal as rhodium, for example, is used in an organometallic catalyst, eseentially complete recovery of the catalyst must be accomplished in order to make the process economical [62].

Over the last five to ten years there has been an increasing effort to solve the separation problem by the reaction of the organometallic complex with a solid containing a ligand that is capable of binding to the complex. Both organic and inorganic substrates have been studied, in particular cross-linked polystyrene and silica [63,64]. The resulting anchored complex retains many of the important features of the homogeneous species. The carbonyl stretching frequencies in carbonyl complexes, for example, are only slightly shifted [65], and the catalytic activity and selectivity patterns of many reactions remain the same [64]. However, this area is still in the early stages of development, and many problems, including complex cross-linking, have to be resolved.

The most promising application for organometallic catalysts is the synthesis of chemicals and clean fuels from synthesis gas. The reactions in these conversions are highly exothermic, and the possibility of liquid phase catalysis offers a very efficient means of temperature control. There have been suggestions that homogeneous catalysts can be designed to be sulfur tolerant [65], but the only indication for such an effect is a Russian report on the hydrogenolysis of carbon–sulfur bonds [66].

It should be recognized that anchoring impairs the stability and the effective heat-transfer capability of the catalyst, especially in gas reactions. Thus, the advantage of anchored catalysts would come in only in high catalyst utilization (effectively atomic dispersion) and unique reaction patterns due to the effect of the ligands. The latter has found an interesting application in stereospecific hydrogenation reactions [67].

9.7 MOLTEN SALTS

A major problem in direct catalytic liquefaction of coal concerns achieving optimum contact between solid catalyst, gas, and liquid phases. Homogeneous catalysis has been explored with limited success in solving this problem. An alternative method is to disperse coal in a catalytic melt involving molten salts. The properties of molten salts have been discussed elsewhere [68,69]. In this section, a brief review of the properties and possible relevance of molten salts to catalysis is presented.

Upon melting, ionic salts are thought to dissociate into ions almost completely. As a result of coulombic interactions [70], each ion in the melt is surrounded by a number of oppositely charged ions. The original long-range order in the salt is destroyed. Molten salts can be best described by a defect solid model known as the hole model. Experimental data seem to support the hole model [69] which attributes an increase in volume from solid to fused states to the empty volumes in the salts.

The species present in melts vary for different salts. A melt of BeF_2, for example, will exhibit polymeric chains while $ZnCl_2$ forms two-dimensional cross-linked layers.

Fused salts have a high capacity for solvating many materials. Gases often dissolve by either reacting or simply filling the free spaces in the melt (hole model). Containment of molten salts is a problem since they dissolve many other inorganic salts as well as most refractory materials. In alkali metal hydroxide melts, the presence of oxygen or water leads to the formation of peroxides which dissolve both noble metals and ceramics. In many instances fused salts also have the ability to dissolve the parent metal. The metal is highly dispersed throughout the medium, and produces metal-like properties such as increased electrical conductivity [71]. Water readily dissolves in many molten salts, especially halides. As an example, $ZnCl_2$ retains some H_2O even at 1000°C [68].

The possible compositions of melts are extensive. Variations in composition give rise to systems with different melting points and very different chemical properties ranging from strong oxidizing to strong reducing, and including a wide range of acidities. This variation offers much flexibility in controlling chemical reaction conditions.

In chemical reactions involving molten salts, the melt can be considered either as a reactant, a solvent, or as a catalyst. Molten salts are especially interesting in catalytic reactions since they can disperse solid reactants for better contacting, remove heat because of their high thermal conductivity, continuously expose fresh catalyst surface, and affect catalytic activity and selectivity by their high polarization forces.

Melts have been used only in a few areas. Chlorination and oxidative chlorination of hydrocarbons such as methane, ethylene, and benzene have been explored using $NaCl/AlCl_3$, $NaCl/AlCl_3/FeCl_3$, and $CuCl_2$ melts [69]. Other work has involved Friedel–Crafts catalytic melts such as $AlCl_3$ and $SbCl_3$ [72]. Recently, Shell Oil Company has described the use of zinc halide melts as hydroconversion catalysts for heavy petroleum fractions [73,74].

Two types of processes have been used in the synthetic fuels area. The first involves coal gasification using alkali metal carbonate or oxide melts at high temperatures such as the use of Na_2CO_3 in the Kellogg Process [75]. The second is demonstrated by the use of $ZnCl_2$ and other Lewis acid halides in the Consol Process of Consolidation Coal Company [76] for hydrocracking, desulfurization, and denitrogenation of coal and coal extract to low sulfur distillate fuel oil or high-octane gasoline. There are several problems with the Consol process, including high catalyst to coal ratios and the necessity of moving large volumes of corrosive catalyst, deactivation of the catalyst, and carbon residue which have to be solved.

9.8 IMPLICATIONS FOR COAL CONVERSION

The discipline of inorganic chemistry impacts in a large number of aspects of catalysis, from the synthesis of catalytic materials to the elucidation of the effect of solid-state and surface chemistry on activity and selectivity. The innovations in inorganic chemistry may have a direct impact in three areas of catalysis concerned with coal conversion: thermal stability, chemical stability, and the more general area of unique and unusual compositions.

Thermal stability is a problem in several highly exothermic coal conversion reactions such as methanation, methanol synthesis, and Fischer–Tropsch reaction, and in the oxidative regeneration of catalysts in liquefaction and many upgrading reactions. This chapter has discussed a number of materials with unusual properties that could be exploited in the development of more stable catalysts. Zeolites represent one group of such materials. Thus zeolites are part of the catalyst system used by Mobil Oil Company to convert CO and H_2 to gasoline [77,78]. The development of ultrastable zeolites discussed in Section 9.2 is an important advance in inorganic chemistry that will have an effect on the expanding use of these materials as catalysts in upgrading processes and catalyst supports in other reactions. The high thermal stability of carbides, nitrides, borides, and silicides also offers an opportunity to improve existing catalysts, especially in view of the potential chemical stability of these compounds. The same applies to the extremely refractory materials, exemplified by $ZrPt_3$, which form by the highly exothermic reaction of transition metals on the right of the periodic table with metals on the left [79].

In coal conversion and the upgrading of coal liquids there are a number of poisons and agents which may affect the chemical integrity of the catalyst. Of particular concern are sulfur, oxygen, and carbon.

In coal liquefaction sulfur is present at high concentrations ($\sim$ 1%) at which only sulfides and certain oxides are sufficiently stable to be used as catalysts. Thermodynamic calculations indicate that borides, silicides, and phosphides of the Group VIII transition metals may be able to resist such an environment as well [1]. This presents some interesting possibilities for liquefaction, and even more so for the catalytic upgrading of gasification products which is particularly sensitive to very small amounts of sulfur. For methanation, for example, the sulfur tolerance of currently used catalysts is less than 10 ppb. Improvements of only one order of magnitude in the sulfur tolerance may have a significant effect on the economics.

Other catalytic materials of potential applicability at low sulfur levels include transition metal exchanged zeolites in which the metal has an electron deficiency and possibly improved sulfur tolerance, alloys, multimetallic clusters in organometallic complexes, and very strong intermetallic compounds such as $ZrPt_3$.

Thermodynamic considerations show that with the exception of oxides most materials are unstable in oxygen at high temperatures. Kinetic limitations may prevent oxidation at lower temperatures, and diffusion may inhibit bulk oxidation for many materials. The use of BaO to stabilize the volatile oxide of Ru in oxidative environment provides an excellent illustration of the application of inorganic chemistry to stability problems in catalysis. Similar use of oxides to stabilize catalytic materials in oxidizing environments against sintering will be discussed in Section 10.4.

In addition to deposition on the catalyst surface, carbon presents another problem to many coal conversion processes by carbide formation. This is particularly important in the case of methanation, where carbiding ranks, together with sintering and sulfur poisoning, as one of the most severe complications [80]. Data on heat of formation indicate [1] that only the nitrides and oxides of most transition metals are more stable than their carbides. However, as one proceeds to the right of the periodic table, carbide stability decreases markedly. Most of the Group VIII carbides are unstable in sulfiding and/or oxidizing environments. Group VIII borides, silicides, and phosphides, on the other hand, are stable [1]. These compounds therefore may possess, in addition to sulfur tolerance, also resistance to carbide formation.

Of the novel structures and compositions mentioned in this chapter only a small fraction has been tested for catalytic applications. Some examples of interesting compositions are worth noting.

The dramatic changes that have been reported in activity and selectivity patterns of reactions upon cluster formation in the Os–Cu and Ni–Cu systems of Sinfelt [81] have implications in methanation, Fischer–Tropsch synthesis, and upgrading of coal liquids. The application of this concept in the synthesis of complex oxides with clusters of transition metals may lead to materials with properties intermediate between those of oxides and of metals [11]. Such materials may be particularly important for reactions which require activation of hydrogen in the presence of carbon monoxide which often interacts too strongly with zero-valent metals to be reactive. Similarly, a metal cluster that maintains its metal-like integrity in an oxidizing environment or in the presence of water is likely to show interesting catalytic properties, especially in the case of multimetallic clusters, in which the various components of the cluster can be utilized. In view of the unique catalytic activity of organometallic clusters not found in their monometallic counterparts and exemplified by the cyclization of acetylene and butadiene in the presence of $Ni_4[CNC(CH_3)_3]_7$ [61], Norton suggested [54] that such cluster complexes may in fact be able to catalyze the reduction of CO by hydrogen. This reaction has not been catalyzed by a mono-organometallic complex to date, but a multimetallic system may aid in achieving the correct electronic balance for its catalysis.

As pointed out in Section 9.3, complex oxides are not the only compounds that exhibit cluster formations with very close metal–metal distances. A similar structure is found in a number of other compounds such as $Al_{0.5}Mo_2S_4$, $PbMo_6S_8$, MoN, Mo_3Se_4 [1], and very likely also in certain thiospinels. Stability in H_2S and the higher hydrogenation–dehydrogenation activity of such cluster-forming compounds can have a significant impact on coal liquefaction and other HDS operations.

Pt-exchanged Y zeolites show much higher hydrogenation activity than expected for Pt[22]. This activity, more characteristic of neighboring Ir than of Pt, has been attributed to electron deficiency [23]. WC has an activity pattern [82] in the reduction of surface oxygen with H_2 and in certain isomerization reactions, which is more typical of a noble metal than of tungsten. The electronic interpretation of this phenomenon is still a controversial subject [83].

Another interesting family of compounds deserving exploration in catalysis is that of the Nowotny carbides and nitrides. They are complex compounds which contain two or more transition metals such as Pt_3ZnC. Several of the members of this class of compounds exhibit a structural similarity to the oxide perovskites mentioned in Section 9.1. However, instead of the oxygen atom being the major component of the structural framework (such as in $BaRuO_3$), the Nowotny perovskites have a metal framework such as $ZnPt_3$ in Pt_3ZnC. The presence of interstitial carbon is likely to strengthen the compound without severly affecting the properties of the metals. Application of such compounds to upgrading of coal liquids and to reactions where carbiding is a problem (such as methanation) would be of interest.

REFERENCES

1. Boudart, M., Cusumano, J. A., and Levy, R. B., New Catalytic Materials for the Liquefaction of Coal, EPRI Rep. RP-415-1, Electric Power Research Institute, October 30, 1975.
2. Schafer, H., "Chemical Transport Reactions." Academic Press, New York, 1964.
3. Anderson, R. B., *in* "Catalysis" (P. H. Emmett, ed.), Vol. 4, p. 1. VanNostrand–Reinhold, Princeton, New Jersey, 1956.
4. Goldstein, M. S., *in* "Experimental Methods in Catalytic Research" (R. B. Anderson, ed.), Chapter 9. Academic Press, New York, 1968.
5. Tanabe, K., "Solid Acids and Bases." Academic Press, New York, 1970.
6. Donnet, J. B., *Bull. Soc. Chim. Fr.* 3353 (1970).
7. Forni, L., *Catal. Rev.* **8**, 65 (1973).
8. Boreskov, G. K., *Adv. Catal.* **15**, 285 (1964).

9. Haber, J., *Int. Chem. Eng.* **15**, 21 (1975).
10. Dickson, J. G., Katt, L., and Ward, R., *J. Am. Chem. Soc.* **83**, 3026 (1961).
11. Tauster, S. J., *J. Catal.* **26**, 487 (1972).
12. Shelef, M., and Gandhi, H. S., *Ind. Eng. Chem. Prod. Des. Dev.* 393 (1972).
13. Shelef, M., and Gandhi, H. S., *Platinum Met. Rev.* **18**, 2 (1974).
14. Wells, A. F., "Structural Inorganic Chemistry," p. 828. Oxford Univ. Press, London and New York, 1975.
15. Eastwood, S. C., Drew, R. D., and Hartzell, F. D., *Oil Gas J.* **60**, 152 (1962).
16. Breck, D. W., *J. Chem. Ed.* **41**, 678 (1964).
17. Lerat, L., Poncelet, G., Dubru, M. L., and Fripiat, J. J., *J. Catal.* **37**, 396 (1975).
18. Kerr, G. T., *J. Phys. Chem.* **73**, 2780 (1969).
19. Kerr, G. T., *J. Phys. Chem.* **71**, 4155 (1967).
20. Hirschler, A. E., *J. Catal.* **2**, 428 (1963).
21. Otnouma, H., Aria, Y., and Ukihaski, H., *Bull. Chem. Soc. Jpn.* **42**, 2449 (1969).
22. Dalla Betta, R. A., and Boudart, M., *in* "Catalysis" (J. Hightower, ed.), p. 1329. North-Holland Publ., Amsterdam, 1972.
23. Gallezot, T., Alarcon-Diaz, A., Dalmon, J. A., Renouprez, J. R., and Imelik, B., *J. Catal.* **39**, 334 (1975).
24. Datka, J., Gallezot, P., Imelik, B., Mossardier, J., and Primet, M., Presentation at the *Int. Congr. Catal. 6th, London,* July 1976.
25. Weisser, O., and Landa, S., "Sulphide Catalysts, Their Properties and Applications." Pergamon, Oxford, 1973.
26. Voorhoeve, R. J. H., and Stuiver, J. C. M., *J. Catal.* **23**, 228 (1971).
27. Vandenberg, J. M., and Brasen, D. J., *Solid State Chem.* **14**, 203 (1975).
28. Omloo, W. P. F. A. M. *et al.*, *Phys. Status Solidi (a)* **5**, 349 (1971).
29. Collin, G. *et al.*, *Nat. Bur. Std. Spec. Publ. 364, Proc. Mater. Res. Symp. 5th* 645 (1972).
30. Biltz, W., and Kocher, A., *Z. Anorg. Allgem. Chem.* **241**, 324 (1939).
31. Tudo, J., and Tridot, G., *C. R. Acad. Sci. Paris* **t258**, 6437 (1964).
32. Tudo, J., and Tridot, G., *C. R. Acad. Sci. Paris* **257**, 3602, (1963).
33. Fransen, F., and Westman, S., *Acta Chem. Scand.* **17**, 2353 (1963).
34. Gleim, W. T., U.S. Patent No. 3,694,352 (1973).
35. Schuit, G. C. A., and Gates, C., *AIChE J.* **19**, 417 (1973).
36. Chevreton, M., and Sapet, A., *C. R. Acad. Sci. Paris* **261**, 928 (1965).
37. Van Laar, B., and Ijdo, D. J. W., *J. Solid State Chem.* **3,** 590 (1971).
38. Yamoka, S., *J. Am. Ceram. Soc.* 111 (1972).
39. Collin, G., and Flahaut, J., *J. Solid State Chem.* **9**, 352 (1974).
40. Barz, H., *Mater. Res. Bull.* **8**, 983 (1973).
41. Brasen, D., Vandenberg, J. M., Robbins, M., Willens, R. H., Reed, W. A., Sherwood, R. C., and Pinder, X. J., *J. Solid State Chem.* **13,** 298 (1975).
42. Levy, R. B., *in* "Advanced Materials in Catalysis" (J. J. Burton and R. L. Garten, eds.). Academic Press, New York, 1977.
43. Toth, L. E., "Transition Metal Carbides and Nitrides." Academic Press, New York, 1971.
44. Bohm, H., *Electrochim. Acta* **15**, 1273 (1970).
45. Wells, A. F., "Structural Inorganic Chemistry," p. 761. Oxford Univ. Press, London and New York, 1975.
46. Barbee, T. W., Jr., Stanford Univ., private communication, 1975.
47. Aronsson, B., Lundstrom, T., and Rundquist, S., "Borides, Silicides, and Phosphides." Wiley, New York, 1965.
48. Searcy, A. W., *in* "Chemical and Mechanical Behavior of Inorganic Materials" (A. W. Searcy, D. V. Ragone, and U. Colombo, eds.), p. 1. Wiley (Interscience), New York, 1970.

49. Nowotny, H., *in* "MTP International Review of Science" (F. R. S. Emelens and L. E. J. Roberts, eds.), Vol. 10, p. 184. Butterworths, London, 1972.
50. Wells, A. F., "Structural Inorganic Chemistry," p. 845. Oxford Univ. Press, London and New York, 1975.
51. Forster, D., and Roth, J. F. (eds.), "Homogeneous Catalysis-II." American Chemical Society, Washington, D.C., 1974.
52. Schranzer, G. N. (ed.), "Transition Metals in Homogeneous Catalysis." Dekker, New York, 1971.
53. Chatt, J., and Halpern, J., *in* "Catalysis, Progress in Research" (F. Basolo and R. L. Burwell, Jr., eds.), Plenum, New York, 1973.
54. Norton, J. R., Catalysis by Polynuclear Complexes: A Bridge Between Homogeneous and Heterogeneous Catalysis, presented at *Am. Chem. Soc. Pet. Chem. Award Symp., New York* April 1976.
55. New Routes to Novel Organic Compounds, *C & E News*, p. 26, January 20, 1975.
56. Karel, K. J., and Norton, J. R., *J. Am. Chem. Soc.* **96**, 6812 (1974).
57. Cariati, R., and Naldini, L., *Inorg. Chim. Acta* **5**, 172 (1971).
58. Albans, V. G. *et al.*, *Chem. Commun.* 859 (1975).
59. Canty, A. J., Johnson, B. F. G., Lewis, J., and Norton, J. R., *Chem. Commun.* 79 (1973).
60. Smith, G. C., Chojnacki, J. P., Dasgupta, S. R., Iwatate, K., and Watters, K., *Inorg. Chem.* **14**, 1419 (1975).
61. Day, V. W., Day, R. O., Kristoff, J. S., Hirsedorn, F. J., and Meutterties, E. L., *J. Am. Chem. Soc.* **97**, 2571 (1975).
62. Roth, J. F., Craddock, J. H., Hershman, A., and Paulik, F. E., *Chem. Technol.* 600 (1971).
63. Allum, J. G. *et al.*, *J. Org. Chem.* **87**, 187, 203 (1975).
64. Bayer, J. C., Jr., *Catal. Rev.* **10**, 17 (1974).
65. NSF Workshop on Fundamental Research in Homogeneous Catalysis as Related To U.S. Energy Problems, held at Stanford Univ., December 4–6, 1974.
66. Berenblyum, A. S. *et al.*, *Izv. Akad. Nauk SSSR Ser. Khim.* **11**, 2650 (1973).
67. Knowles, W. S., Sabacky, M. J., and Vineyard, B. D., *in* "Homogeneous Catalysis-II" (D. Forster and J. F. Roth, eds.), p. 274. American Chemical Society, Washington, D.C., 1974.
68. Bloom, H., "The Chemistry of Molten Salts." Benjamin, New York, 1967.
69. Sundermeyer, W., *Angew. Chem. Int. Ed.* **4**, 222 (1965).
70. Temkin, M., *Acta Physicochim. USSR* **20**, 411 (1945).
71. Bronstein, H. R., and Bredig, M. A., *J. Am. Chem. Soc.* **80**, 2077 (1958).
72. Cheney, H. A., U.S. Patent No. 2,342,073 (1944).
73. Hardesty, D. E., and Rodgers, T. A., U.S. Patent No. 3,677,932 (1972).
74. Geymer, D. O., U.S. Patent No. 3,844, 928 (1974).
75. Cover, A. E., Schreiner, W. C., and Skaperdas, G. T., *Chem. Eng. Progr.* **69**, 31 (1973).
76. Consolidation Coal Company, R & D Rep. No. 39, OCR Contract No. 14-01-0001-310, Vols. I, II, III (1969).
77. Chang, D., Silvestri, A. J., and Smith, R. L., Mobil Oil Corp., U.S. Patent No. 3,894,103 (July 8, 1975); U.S. Patent No. 3,894,104 (July 8, 1975).
78. Chang, C. D., and Silvestri, Mobil Oil Corp., U.S. Patent No. 3,894,102 (July 8, 1975).
79. Brewer, L., *Acta Metall.* **15**, 553 (1967).
80. Mills, G. A., and Steffgen, F. W., *Catal. Rev.* **8**, 159 (1973).
81. Sinfelt, J. H., *J. Catal.* **29**, 308 (1973).
82. Levy, R. B., and Boudart, M., *Science* **181**, 547 (1973).
83. Colton, R. J., Huang, J. J., and Rabalais, J. W., *Chem. Phys. Lett.* **34**, 337 (1975).

Chapter 10

Materials Science

Materials science is a broad field, encompassing a number of areas of chemistry, physics, and engineering, including solid-state physics, inorganic chemistry, and ceramics engineering. An important aspect of this field is the synthesis and characterization of materials which have unusual mechanical and thermal properties. This chapter will review four areas: monolithic supports, intermetallic compounds, novel inorganic polymers, and sintering.

10.1 MONOLITHIC SUPPORTS

The development of monolithic structures stemmed from the need for thermally, mechanically, and chemically durable supports for automotive emissions control catalysts, and resulted in a commercial technology capable of producing materials with a wide range of properties suitable for other applications in catalysis.

Monolithic supports contain small-size parallel channels of a variety of shapes and diameters. These structures may be in the form of ''honeycomb'' ceramics extruded in one piece, oxidized aluminum alloys in rigid cellular configurations, or multilayered ceramics formed by corrugation procedures. The channels in honeycomblike structures commonly have tubular diameters of 1–5 mm. The overall diameter of monolithic supports may vary from 1 in. to 2 ft. Materials of fabrication are usually low surface area ceramics such as mullite ($3Al_2O_3 \cdot 2SiO_2$) or cordierite ($2MgO \cdot 5SiO_2 \cdot 2Al_2O_3$). More recently, silicon carbide, silicon nitride, and zirconia have also been fabricated into monoliths. The refractory monolith is produced with macro pores (1–10 μm), and is usually provided with a 5–20 wt% γ-alumina washcoat which serves as support for the precious metal catalyst. Two major advantages of monolithic supports are high superficial or geometric surface area and low pressure drops during operation. In comparing monoliths with typical packed bed systems, it is not unusual to observe a pressure drop decrease of over one order of magnitude for the same geometric surface area [1]. In addition, many of the monolith materials have good thermal conductivity which is an added asset for use as catalyst supports for highly exothermic reactions. A typical comparison of the properties of a monolith versus a packed bed is shown in Table 10-1 for the Engelhard HC-series monolith catalysts [1].

A number of manufacturers fabricate a total of almost 8 million monoliths annually for automotive use. Corning's Celcor is an example of a cordierite honeycomb structure produced by an extrusion process with square and triangular cell shapes. Dupont's Torvex is prepared by controlled oxidation of machined aluminum alloys, and is available in alumina or mullite. It is made in three different orientations: a conventional straight cell, a cell slanting at 45° from the open cell face, and a crossflow configuration consisting of half cells angled 90° to each other and 45° to the cell face. American Lava's Thermacomb, a corrugated cordierite ceramic, is manufactured in two basic structures, honeycomb (no separator between corrugated layers) and split-cell (with separator). From these two basic structures, three additional configurations can be formed, crossflow split-cell (alternate cell layers at 90°), crisscross split-cell, and crisscross honeycomb (alternate cell layers at 45°). Other manufacturers include General Refractories and Davison Division of W. R. Grace.

Cordierite and mullite are the common materials of construction for commercial monoliths. The mullite structures are more sensitive to thermal shock, but are more stable at higher temperatures. Cordierite structures are stable up to 2200°F, above which gradual degradation occurs resulting in the precipitation of mullite [2,3] and the formation of spinel according to

$$\underset{\text{cordierite}}{2(2MgO \cdot 2Al_2O_3 \cdot 5SiO_2)} + \underset{\alpha\text{-alumina}}{15Al_2O_3} \rightarrow \underset{\text{mullite}}{5(3Al_2O_3 \cdot 2SiO_2)} + \underset{\text{spinel}}{4(MgO \cdot Al_2O_3)}$$

The α-alumina is a conversion product of the γ-alumina in the washcoat.

For automotive emissions control the catalyst is typically 0.3 wt% Pt or Pt–Pd on a cordierite monolith which contains a 10–15 wt% washcoat of Al_2O_3 [2–11]. The primary advantages of these structures over normal packed beds are low pressure drop and minimum attrition rates. This is particularly important in view of the high gas flows (100,000–200,000 V/V/hr) and substantial mechanical

TABLE 10-1

COMPARISON OF PROPERTIES OF HC-CATALYSTS WITH PARTICULATE TYPES[a]

	HC-series		Particulate types	
	7 cpi[b]	$4\frac{1}{2}$ cpi[b]	$\frac{1}{8}$-in. particles	$\frac{1}{4}$-in. particles
Surface area (m^2 $liter^{-1}$)				
Superficial	1.98	1.17	1.11	0.48
Catalytic	4×10^5	4×10^5	1.28×10^5	3.44×10^5
Voids in packed bed (%)	65–70	82	38.5	41.5
Pressure drop per foot of depth at linear velocity of 20 ft/sec, pressure of 100 psig, temperature of 500°C	0.6 psi	0.3 psi	16.5 psi	10.1 psi
Ratio of vessel diameters for ΔP of 1 psi at operating conditions (100,000 space velocity for HC, 60,000 for particles referred to 1.0 for HC-4 $\frac{1}{2}$ cpi particles)	1.25	1.0	4.0	3.25
Commercially usable space velocities (SCFH/CF)	90,000–140,000	90,000–140,000	20,000–60,000	20,000–60,000
Bed orientation limits	None, can be mounted horizontally, vertically, etc.		In general, catalyst must be contained in vertical column, with gas overflow	

[a]From Andersen *et al.*, [1], courtesy of Engelhard Minerals and Chemicals.

[b]Channels per linear inch.

vibrations occurring in this application. The catalysts operate at 1300–1600°F, and survive frequent transients up to 2000°F. The catalysts are reported to be effective for 50,000 miles in controlling hydrocarbons and CO within EPA specifications. Monolith supports have also been studied for cumene cracking and ethanol dehydration by Campbell [12]. He showed that the monolithic geometry of Corning's extruded structures can give higher rates than a packed bed, and also improve a low effectiveness factor, which is the fraction of the internal surface of the catalyst utilized.

Other applications for these structures include NO_x abatement [1], catalytic combustion for energy generation [13,14], and emissions control for incinerators [15].

In the area of the catalytic conversion of coal, monolithic materials might be considered for application in methanation, Fischer–Tropsch synthesis, and the water–gas shift reaction. Their use will depend significantly on obtaining marked advantages in catalyst life and activity over an equivalent volume of packed bed catalyst in order to justify the higher cost of monolithic supports.

10.2 INTERMETALLIC COMPOUNDS

An intermetallic compound is an alloy of two or more metals which has a characteristic crystal structure and a definite stoichiometry. This is in contradistinction to homogeneous alloys for which no order may be present. A group of binary intermetallic compounds exhibits remarkable chemical and thermal stability and contains metal–metal bonds which are more directed, and in many instances stronger, than those characteristic of substitutional or interstitial alloys. These intermetallic compounds usually consist of a combination of two metals from opposite sides of the periodic table and have been extensively investigated by metallurgists.

10.2.1 Properties of Intermetallic Compounds

Much of the work in this area has been stimulated by the studies of Engel [16–19], Brewer [20–23], and Hume-Rothery [24–27]. The work of Engel and Brewer has resulted in the development of correlations which predict the structures of alloys and intermetallic compounds.

In using the Engel–Brewer correlation, it is assumed that the crystal structures of the metallic elements are based on definite electronic states of the composite atoms. The unpaired electrons take part in crystal bonding, but the d electrons play no part in determining the type of crystal structure, which is controlled only by the number of s and p electrons. The bcc, hcp, and fcc structures are regarded as resulting from 1, 2, and 3(s + p) electrons, respectively. However, the d electrons can have a profound effect on those properties which are related to the electronic configuration changes between the pure elements and the intermetallic compound.

For predicting structure or thermodynamic properties one needs to define the electronic state of the atoms in the pure metals and in the intermetallic compound formed from these metals. This is usually done by choosing the ground state of the free atom or the lowest excited state which avoids the existence of s^2 subgroups, since the latter is considered as nonbonding. In some instances, this configuration is rejected in favor of one which involves more bonding electrons. For example, for rhenium, the ground state is d^5s^2, and the first excited state is d^6s (excitation energy of 6 kcal $mole^{-1}$) with a second excited state of d^5sp (excitation energy of 46 kcal $mole^{-1}$). Here the d^6s configuration involves 5 bonding electrons because two of the d electrons are paired, and therefore it is rejected in favor of the d^5sp configuration which has seven bonding electrons and 2(s + p) electrons. Thus the rule predicts that rhenium forms an hcp structure, which is in fact the case.

As another example, the ground state configuration of Mg is sp, and this indeed agrees with the observed hcp structure for the metal. Aluminum has an sp^2 electron configuration which also explains its fcc structure.

The Engel–Brewer correlations may be applied to alloys in a similar way. Here, Brewer has determined the tolerance one might expect on either side of the 1, 2, and 3(s + p) electrons which correspond to the bcc, hcp, and fcc structures, respectively. Having determined these limits, the corresponding compositions in a given binary alloy or intermetallic system can be estimated, and therefore the phase diagram determined. This has been done successfully for a number of intermetallic systems [20–22].

The Engel–Brewer correlation also provides a means for calculating thermodynamic properties provided that the electronic configuration of the gaseous elements is known. This correlation predicted the existence of the so-called Brewer compounds [20–22] and led to the experimental demonstration of their extraordinary thermal and chemical stabilities [21,28]. Thus the application of the Engel–Brewer correlations to transition metal alloys predicts that intermetallic compounds of metals of the second and third transition series prepared by combining transition metals from the right and left columns of the periodic table will have unusual stability. By this model, transition metals of the fourth and fifth periods and Groups I–VII, i.e., Rb to Tc and Cs to Re, use all of their valence electrons for bonding, while Groups VIII–IB, i.e., Ru to Ag and Os to Au, do not use all of their valence electrons for bonding in the pure metals. This is because some of the d electrons are paired internally and are unavailable for bonding. As Brewer indicates [21], the difference between the metals from the left and right sides of the periodic table can be summarized in terms of incomplete use of available bonding orbitals and complete use of available electrons on the left side (e.g., Zr), and incomplete use of electrons and full use of bonding orbitals on the right (e.g., Pt). For an intermetallic compound formed from these two classes of metals, an electron transfer occurs from the metal with a surplus of unpaired electrons to the metal with low-lying vacant bonding orbitals to maximize utilization of all electrons and orbitals. This gives rise to unusual thermal and chemical stability.

The formation of $ZrIr_3$ is an example of this stabilization. According to the Engel–Brewer correlation, the electron configuration of Zr in the hcp structure is d^2sp. The configuration of Ir in the fcc structure is d^6sp^2 with only seven electrons available for bonding since one pair of d electrons is paired internally. In forming $ZrIr_3$, Ir donates electrons to Zr to approach the configuration d^5sp^2, while Zr accepts these electrons to assume the configuration d^4sp^2. If these configurations are attained, the number of electrons used for bonding in the intermetallic compound is increased from 25 for the pure elements to as many as 31 per $ZrIr_3$ entity in the intermetallic compound.

For a given element on the left side of the periodic table, e.g., Zr, the number of electrons transferred, and therefore the increase in the number of bonding electrons is predicted to increase as the other metal is changed from Os to Pt. The stability of the intermetallic compounds should therefore increase in the same direction; i.e.,

$$\text{Zr–Pt} > \text{Zr–Ir} > \text{Zr–Os}$$

Brewer therefore predicts that for a given metal on the left, say Zr, the stability of the intermetallic compound should reach a maximum at Group VIII_3 (Ni, Pd, Pt). Similarly, for Pt in combination with metals on the left, say Mo, Nb, Zr, stability is predicted to go through a maximum at Group IVB (Ti, Zr, Hf).

In spite of the controversy associated with the broad application of the Engel–Brewer correlations, in a number of instances experimental data substantiate the qualitative, and in many cases the quantitative, aspects of these correlations.

The stability of Brewer compounds $ZrPt_3$ and $ZrIr_3$ is shown by experiments in which Pt, Ir, and Os were heated with ZrC [21]. In each case, the carbide is decomposed to form graphite and the corresponding intermetallic compound. Accordingly these intermetallic compounds have free energies of formation which are more negative than -47 kcal $mole^{-1}$ of ZrC, one of the most stable carbides known. The compounds formed were identified as hcp $ZrPt_3$ (ordered $TiNi_3$ structure), $ZrIr_3$ (ordered $AuCu_3$ structure), and hcp solution of Zr in Os.

Another impressive example is the work reported by Brewer [21] in which a ZrO_2/Pt mixture was reduced in hydrogen, giving $ZrPt_3$ at only 1200°C. The enthalpies of such reactions are so negative that in some instances explosions occurred in the adiabatic reaction between the powdered elements initiated at 1000°C [29].

Ficalova *et al.* [30] investigated the thermal and chemical stability of 7000 Å thick films of $HfPt_3$ and $ZrPt_3$ prepared from the elements, by subjecting them and similar Pt films to 60-sec pulses of oxypropane and oxyacetylene flames. The Pt film failed while there was no effect on the intermetallic compounds. Similarly exposure to boiling aqua regia showed the intermetallics to have superior chemical resistance.

Berndt *et al.* [31] was able to prepare Rh_3Th, Rh_3U, Pt_5Th, and other intermetallic compounds from the respective oxides and noble metals at relatively low temperatures in the presence of hydrogen because of the high thermodynamic driving force provided by the enthalpy of formation for the intermetallic compound. This is remarkable in view of the almost impossible task of reducing ThO_2 or UO_2 with hydrogen to the metal at even the highest attainable temperatures. Similar results were observed by Darling *et al.* [32,33] for Pt/ZrO_2 and Pt/ThO_2, and confirmed by Erdmann and Keller [34]. In the latter study, Rh_3Th, Rh_3U, Pd_3Th, Pd_4U, Ir_3U, Pt_3U, Pt_3Th, and other intermetallic compounds were formed in hydrogen from the oxide and noble metal at temperatures as low as 1200°C. Further studies [35] showed that numerous ultrastable intermetallic compounds could be formed between the Group VIII metals and the rare earths and actinides.

10.2.2 Catalytic Applications

The problem connected with the catalytic use of the unusually stable intermetallic compounds is the difficulty in preparing these materials in a high surface area form. One way to overcome this problem in the case of $ZrPt_3$, for example, would be to impregnate ZrO_2 or an Al_2O_3-supported ZrO_2 with a Pt salt, and then to reduce in very dry pure hydrogen. Because the thermodynamic driving force to form this compound is of the order of $\Delta H \gtrsim 100$ kcal mole^{-1}, the temperature required to form it is substantially lower than necessary to form other conventional alloys with Group IVB metals, but still too high (~1100–1200°C) for maintaining high surface area. Thus, ZrO_2 and Al_2O_3 would be expected to sinter to surface areas less than 1 m^2 gm^{-1}. To counteract this effect one would have to resort to the use of thermally treated or chemically stabilized aluminas, or of ultrastable zeolites when preparing Brewer compounds.

There have been only a few catalytic studies with the Brewer compounds. In fuel cell catalysis Russian workers have used $ZrAl_3$ and other related compounds prepared by electrodeposition [36]. In a similar application Ewe *et al.* [37] studied $LiNi_5$ as an oxidation catalyst for a hydrogen fuel cell and obtained encouraging activity.

Some more easily prepared intermetallic compounds such as Cu_3Au, Cu_2Mg, AgSb, SbSn, Cu_2Sb, and WCo_3 [38] have been examined as catalysts for simple hydrogenation reactions and formic acid decomposition. While various electronic theories have been developed to explain the catalytic data and their correlation with the properties of the intermetallic compounds, the results have been relatively unimpressive.

A patent issued to the British Petroleum Company [39] discloses the use of numerous intermetallic compounds such as Pt_3Co, $PtNi_3$, $PtCu_3$, and others for catalytic reforming of naphthas to high-octane gasoline. A number of these intermetallic compounds were supported on Al_2O_3, and claimed to be excellent catalysts for the conversion of paraffins to aromatics.

10.2.3 Future Directions

The broad class of intermetallic compounds, especially the very stable Brewer compounds, offers a number of potential applications in catalysis, possibly in the synthetic fuel area. Thus, it may be possible to prepare supported intermetallics, which have unusual thermal and chemical resistance, by combinations of the Group VIII metals with metals like Ti, Zr, Hf, V, Nb, Ta, Cr, Mo, and W. Because of the very strong interactions involved in forming these compounds, sintering is expected to be significantly reduced. This strong interaction is likely to modify the electronic and therefore the catalytic properties of the Group VIII metal. In some instances, this may result in less desirable catalytic properties. For example, in the case of $ZrPt_3$ the extensive electron withdrawal by the Zr atoms renders the Pt highly electron deficient, and therefore less metallic than zero-valent platinum. It may be possible to decrease and control this excessive interaction by using elements to the left or right of Zr (e.g., Mo) as the second element.

The metal–metal interaction in intermetallic compounds can also be used to modify the catalytic properties of a metal in a manner analogous to the effect observed by Dalla Betta and Boudart [40] for Pt-loaded zeolites in which atomic clusters of Pt were found to donate electrons to the zeolite structure, making the Pt behave catalytically like Ir.

Another property affected by this electron transfer is tolerance to sulfur, nitrogen, arsenic, and phosphorus which poison metals like Pt by adsorbing strongly on their surface with electron withdrawal. If the interaction between Pt, Pd, or any other metal on the right side of the periodic table with one on the left is strong and follows Brewer's prediction of electron transfer, then the interaction of these catalytic metals with the mentioned electrophilic poisons would likely to be substantially reduced.

Intermetallic compounds may find applications in upgrading coal liquids, methanation, water–gas shift catalysis, and Fischer–Tropsch synthesis if ways can be found for their preparation in high surface area form.

10.3 NOVEL INORGANIC POLYMERS

There are a few other classes of materials not yet discussed which offer some possibilities for applications in coal conversion catalysis. The first of these depends on technology to spin refractory oxides into high surface area fibers. For example, Imperial Chemical Industries, Ltd., offers their Saffil Al_2O_3 and ZrO_2 fibers with surface areas varying from 1.5 to 160 m^2 gm^{-1} for Al_2O_3 and 7.5 to 10 m^2 gm^{-1} for ZrO_2 [41]. The Al_2O_3 fibers are reported to withstand temperatures in excess of 2750°F, while the ZrO_2 fibers can be used at 2900°F. The fibers can be fabricated into numerous forms including monolithic structures, thimbles, cylindrical filters, and cloths. Because of the thermal stability, high surface area, and the fabrication properties of these materials, they represent a possible new class of catalyst supports. Potential applications include many of those for which monolithic supports are used. Only minimal catalytic work has been reported as yet for these materials [41], and their high cost militates against a thorough evaluation of their real potential.

A second broad class of materials includes new inorganic polymers of nitrogen with phosphorus [42,43] and of nitrogen with sulfur [44]. Not enough work has been done with the sulfurnitrogen polymers to speculate as to whether these materials might have an application in catalysis. The phosphorus-nitrogen polymers, on the other hand, exhibit interesting properties which are worth noting.

The nitrogen phosphorus polymers, called polyorganophosphazenes, have recently aroused considerable technological interest because of numerous applications ranging from high-temperature lubricants and polymers to new synthetic rubbers.

They contain phosphazene units which are cyclic (a) and (b) or linear (c) molecules containing alternate phosphorus and nitrogen atoms in the skeleton, with two substituent groups attached to each phosphorus [42]:

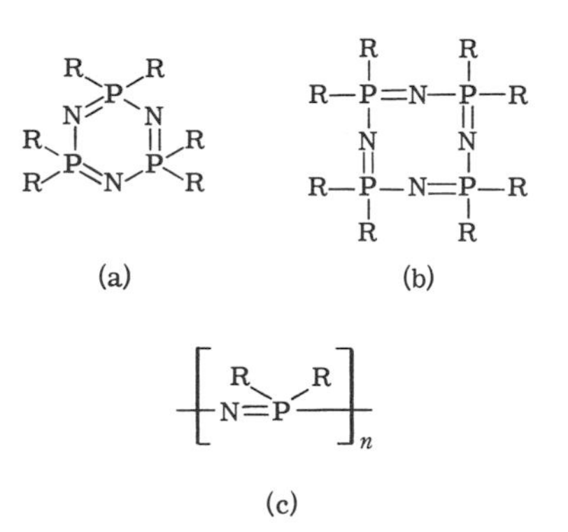

The cyclic trimer and tetrameric chlorophosphazenes have been known for many years. They are synthesized by elimination of HCl from the reaction between NH_4Cl and PCl_5 at about 120°C. They are readily converted, by heating, to a transparent elastomeric polymer, called inorganic rubber [42]. Its elastomeric characteristics persist from −50 to +350°C. It swells in organic solvents, but does not dissolve. It also has remarkable resistance to thermal degradation. Indeed, it would be a valuable technological material were it not for the fact that it slowly reacts with moisture in the atmosphere and hydrolyzes to H_3PO_4, HCl, and NH_3. This hydrolytic decomposition is brought about by reaction of moisture with the chlorine groups and not by scission of nitrogen–phosphorus bonds. To solve this problem, recent studies have been directed at replacing the chlorine with other groups [43,44], and led to the synthesis of a number of these polymeric materials with high thermal and chemical stability.

Possible applications of these polymers in catalysis are primarily as support materials because of their mechanical, thermal, and chemical stability and the possibility of adding catalytic functions by attaching various groups to the polymer, including organometallics.

10.4 SINTERING

Sintering is the coalescence of small particles or atoms at elevated temperatures caused by diffusion and resulting in a loss of surface area. Materials scientists have studied this phenomenon in connection with a number of practical applications. The techniques used in powder metallurgy depend heavily on an understanding of sintering and surface diffusion. The rate of migration and coalescence of internal voids in metals, occurring in nuclear reactors, is governed by surface diffusion processes [45]. The process of crystal growth from the vapor phase has been recognized to depend on surface diffusion of adatoms [46]. The technology of the film semiconductor devices is related to surface diffusion through nucleation and growth of the epitaxial films [47].

More recently, catalysis research has begun to apply the concepts of sintering developed by material science to problems of catalyst deactivation caused by sintering [48–54].

Most of the work on sintering of catalytic interest has been done on metals, and in particular on supported metal catalysts. Ruckenstein and Pulvermacher [55,56] developed a model for metal sintering which involves migration, collision, and fusion of metal crystallites on the support surface. In this treatment, either migration or coalescence can be rate determining, and the rate of change of the metal surface area S is described by an equation of the form

$$dS/dt = -KS^n \tag{10-1}$$

where K is a constant and n can vary from 2 to 3 for diffusion-limited rates, and from 4 to 8 for coalescence-limited rates.

The scheme developed by Flynn and Wanke [54,57] treats the sintering process as a two-dimensional evaporation–condensation process. An equilibrium is assumed to be reached between metal atoms on the metal catalyst particles and those which have migrated to the support surface. The rate at which atoms dissociate from metal particles is assumed to be independent of particle size, while the rate of atom recapture is taken to be proportional to the particle diameter. Therefore, larger particles grow at the expense of smaller ones. In accordance with the Kelvin equation, the concentration of metal atoms in equilibrium with small particles is greater than that around larger particles. The difference in concentration leads to a transport of metal atoms from smaller to larger particles, a phenomenon known for the case of solid particles in a liquid phase as Ostwald ripening. The Flynn–Wanke model suggests that the distribution of particle sizes determines the value of the exponent in Eq. (10-1) and that n will vary during the course of any sintering process.

Wynblatt and Gjostein [52] proposed a model which considers in some detail the various processes involved in the sintering of supported metal catalysts including crystal nucleation and growth, particle migration and coalescence, and surface or vapor transport of the metal. These authors calculate order of magnitude rates for these various steps using data obtained primarily for pure metals or supported metal films rather than supported metal catalysts. They conclude that particle growth can occur in either a noninhibited or inhibited mode. As Schlatter [48] points out, the Ruckenstein–Pulvermacher model (migration, collision, and coalescence) and the Flynn–Wanke model (dissociation, diffusion, and recapture) are examples of noninhibited particle growth. Wynblatt and Gjostein's idea of inhibited growth stems from their observation that platinum particles on a flat alumina surface approach a limiting size. They attribute this to faceting of the metal crystallites. Therefore, crystallite growth is suggested to be inhibited by the requirement of a nucleation process for each new layer of atoms added to a particle.

There are a number of differences among these three models. For example, the Ruckenstein–Pulvermacher model assumes transport by particle migration whereas the Flynn–Wanke model makes use of atomic diffusion. Ruckenstein and Pulvermacher argue that the interaction between metal atoms and common catalyst support surfaces would be too weak to make dissociation energetically favorable as suggested by Flynn and Wanke. They also mention that the rates of atom evaporation would be too low to account for observed sintering rates. According to the Flynn–Wanke and Wynblatt–Gjostein models, the diffusion can be two-dimensional on the surface, or three-dimensional into the gas phase with readsorption on the surface of the support. This is particularly true in the presence of reactive gases such as oxygen, carbon monoxide, or halogens. For example, nickel is readily transported both on the surface and in the gas phase as is nickel carbonyl in the presence of CO. Similarly, platinum is transported as PtO_2 in the presence of oxygen. Reactive gases can, however, promote sintering only at conditions at which the bulk metals are stable. Neither the metal compound formed nor the surface compound with the support can be more stable than the bulk metal; otherwise, increased dispersion of the particles will result.

Wynblatt and Gjostein agree that gas phase or surface migration of metal atoms is ruled out in a reducing environment, but suggest that in an oxidizing environment some metals can migrate as oxides. The differences between these models point to the need for additional data to clarify the nature of the sintering process and to provide ideas for model development and refinement.

In spite of their differences, the three models agree on some general aspects. All three models predict that large metal crystallites will grow at the expense of small ones at elevated temperatures, as is observed experimentally. One might also deduce that the pore structure of the support would tend to stabilize the metal dispersion, as would increased metal–support interactions. These effects are also observed experimentally, and have been discussed in Chapter 3. According to Schlatter [48], all three models can account for the fact that platinum sinters more readily in an oxidizing environment than in a reducing environment. However, this appears to be an after-the-fact correlation with experiments. Indeed, there is a real need for more controlled experimental kinetic data to develop more exact theories of the sintering process.

It is well documented that supported metal catalysts are thermally more stable than unsupported metal catalysts. Thus, while platinum sinters severely under even the mildest conditions [58–60] when unsupported, when dispersed on refractory oxide supports such as Al_2O_3 it is stable even at 800°C [61].

The sintering rate depends markedly on temperature, time, and gaseous environment. Herrmann *et al.* [62] found that the sintering rate for platinum supported on γ-Al_2O_3 increased with increasing temperature and time in nitrogen, and was second order in the platinum area remaining. Maat and Moscou [61] confirmed this second-order behavior for a platinum on γ-Al_2O_3. However, Kirklin and Whyte [63] found that platinum on alumina catalysts do not follow this second-order dependence in a hydrogen. This suggests different mechanisms for the two environments. Somorjai *et al.* [64] reached this same conclusion by studying the sintering of platinum on η-Al_2O_3 in both reducing and oxidizing atmospheres. The sintering rate was found to be faster under oxidizing conditions than under reducing conditions [64–66], having activation energies in the range of 44–55 and 14–27 kcal $mole^{-1}$, respectively.

In intriguing sintering experiments, Emelianova and Hassan [60] found that rapid cooling of a platinum catalyst maintained a high metal surface area. In contrast, slow cooling resulted in a substantial amount of sintering. In view of the above models, one might speculate that slow cooling permits the metal particles which are mobile at these high temperatures to coalesce, and therefore to lose surface area. On the other hand, rapid cooling freezes the mobile and highly dispersed transport species present at high temperatures.

Some sintering studies on supported metal catalysts have been done with rhodium [53], palladium [67,68], and nickel [67,68]. High temperatures caused the silica-alumina-supported Pd particles to increase in size in O_2 while in H_2 only an annealing of the lattice defects took place with a resultant loss in specific catalytic activity, but not in surface area. It was also found that the addition of high-valent cation (e.g., Th^{4+}, La^{3+}) enhanced the thermal stability—an effect possibly due to hindered surface mobility of Pd because of increased metal–support interaction.

Schlatter [48] found that the sintering of Pd on Al_2O_3 in H_2 and O_2 environments differed from the results described above for $SiO_2 \cdot Al_2O_3$-supported Pd. After 4 h at 650°C in O_2 the surface area showed no loss, but decreased by almost 75% after the same treatment in H_2. Even more interestingly, O_2 treatment of the sintered sample at 650°C restored the Pd area. It would appear that the O_2 provides a medium for forming surface PdO species which redisperse the metal on the Al_2O_3 support. These results emphasize the complexity of the sintering process and its dependence on the type of system that is being considered. Indeed, to completely describe the sintering process, one must consider the following variables:

(1) structure and stability of the metal surface,
(2) equilibrium shape or configuration of the metal particles,
(3) stability of metal particles with respect to dissociation as a function of particle size and shape,
(4) mobility of metal atoms over the metal surface,
(5) interaction of metal particles and individual atoms from these particles with the support surface,
(6) mobility of metal particles and atoms over the support surface, and
(7) support morphology.

Before the relative merits of the various theoretical models can be evaluated more extensive experimental data are needed for metal particle distribution as a function of time, temperature, and gaseous environment. New improved electron microscopic techniques capable of recording events on film or video tape in the manner of Baker [69] should be very useful in this respect.

Another area of substantial technological value which warrants further work is the development of techniques for the redispersion of sintered metals. In the case of the Pt on alumina system it was found some time ago that mixtures of chlorine and oxygen could be used to redisperse agglomerated platinum [70]. Similar techniques need to be developed for other metals and particularly for combinations of Group VIII and IB metals for use in coal conversion. Certain upgrading processes discussed in Part III, Fischer–Tropsch synthesis, water–gas shift catalysis, and methanation are additional areas of possible interest.

REFERENCES

1. Andersen, H. C., Romeo, P. L., and Green, W. J., *Engelhard Tech. Bull.* **8**, 100 (1966).
2. Voltz, S. E., and Liederman, D., *Ind. Eng. Chem., Prod. Des. Dev.* **13**, 243 (1974).
3. Mooij, J., Kuebrick, J. P., Johnson, M. F. L., and Chloupek, F. J. paper presented at 74th National Meeting, American Institute of Chemical Engineers, New Orleans, Louisiana, March, 1973.
4. Aykan, K., Hindin, S. G., Kenson, R. E., and Mooney, J. J., paper presented at *Can. Chem. Eng. Conf., 22nd, Toronto, Canada* September, 1972.
5. Aykan, K., Hindin, S. G., and Kenson, R. E., paper presented at *I&EC Div. Symp.* American Chemical Society, Chicago, Illinois, January, 1972.
6. Aykan, K., Mannion, W. A., Mooney, J. J., and Joyer, R. D., paper No. 730592, Automobile Engineering Meeting, Society of Automotive Engineers, Detroit, Michigan, May, 1973.
7. Aykan, K., Mooney, J. J., Hoyer, R. D., and Manion, W. A., paper presented at 38th Midyear Meeting, Division of Refining, American Petroleum Institute, Philadelphia, Pennsylvania, May, 1973.
8. Dwyer, F. G., *Catal. Rev.* **6**, 261 (1972).
9. Liederman, D., Voltz, S. E., and Oleck, S. M., paper presented at Division of Petroleum Chemistry, American Chemical Society, Dallas, Texas, April, 1973.
10. Liederman, D., Voltz, S. E., and Snyder, P. W., paper presented at 3rd North American Meeting, Catalysis Society, San Francisco, California, February, 1974.
11. Shelef, M., Dalla Betta, R. A., Larson, J. A., Otto, K., and Yao, H. C., paper presented at 74th National Meeting, American Institute of Chemical Engineers, New Orleans, Louisiana, March, 1973.
12. Campbell, L. E., paper presented at *Spring Symp. Philadelphia Catal. Club, Philadelphia, Pennsylvania* May 7, 1974.
13. Pefferle, W. C., Carrubba, R. V., Heck, R. M., and Roberts, G. W., paper presented at the Winter Annual Meeting of the Fuels Division of The American Society of Mechanical Engineers, Houston, Texas, November 30–December 4, 1975.
14. Anderson, D. N., Tacina, R. R., and Mroz, T. S., NASA Technical Memorandum, NASA TM X-71747, June, 1975.
15. Searles, R. A., *Chem. Ind.* **22**, 895 (1974).
16. Engel, N., *Ingenioeren* **N101** (1939); **M1** (1940).
17. Engel, N., *Powder Met. Bull.* **7**, 8 (1954).
18. Engel, N., *Am. Soc. Met. Trans. Quart.* **57**, 610 (1964).
19. Engel, N., *Acta Metall.* **15**, 557 (1967).
20. Brewer, L., *Science* **161**, 115 (1968).
21. Brewer, L., *Acta Metall.* **15**, 553 (1967).
22. Brewer, L., *in* "High-Strength Materials" (V. F. Zackay, ed.), pp. 12–103. Wiley, New York, 1965.
23. Brewer, L., *in* "Phase Stability in Metals and Alloys" (P. Rudman, J. Stringer, and R. I. Jaffee, eds.), pp. 39–61, 241–249, 344–346, 560–568. McGraw-Hill, New York, 1967.
24. Hume-Rothery, W., *Acta Metall.* **13**, 1039 (1965).
25. Hume-Rothery, W., "The Metallic State." Oxford Univ. Press, London and New York, 1931.
26. Hume-Rothery, W., "Structures of Metals and Alloys." Institute of Metals, London, 1936.

27. Hume-Rothery, W., *Prog. Mater. Sci.* **13**, No. 5, 229 (1967).
28. Ficalora, P., Wu, J., Srikrishnan, V., and Carloni, J., Final Rep. under NOSC Contract N00017-70-C-4416, March, 1971, Syracuse Univ., Dept. of Chem. Eng. and Matl. Sci., Syracuse, New York.
29. Margrave, J., footnote in Brewer (21).
30. Ficalora, P. J., Srikrishnan, V., and Pecora, L., NOSC Contract N00017-72-C-4424, January, 1974.
31. Berndt, U., Erdmann, B., and Keller, C., *Platinum Met. Rev.* **18** (1), 29 (1974).
32. Darling, A. S., Selman, G. L., and Rushforth, R., *Platinum Met. Rev.* **14**, (2), 54 (1970).
33. Darling, A. S., Selman, G. L., and Rushforth, R., *Platinum Met. Rev.* **14** (3), 95 (1970).
34. Erdmann, B., and Keller, C., *J. Solid State Chem.* **7**, 40 (1973).
35. Erdmann, B., and Keller, C., *Inorg. Chem. Lett.* **7**, 675 (1971).
36. Asanov, U. A., Sakavov, I. E., and Denisov, A. S., *Mekhandemissiya Mekhanokhim. Tverd. Tela.* 173 (1974).
37. Ewe, H., Justi, E. W., and Stephan, K., *Energy Convers.* **13**, 109 (1973).
38. Schwab, G. M., and Ammon, R., *Z. Metallk.* **52**, 583 (1961).
39. South African Patent 71/2027 (1970).
40. Dalla Betta, R. A., and Boudart, M., *in* "Catalysis" (J. W. Hightower, ed.), Vol. 2, p. 1329. American Elsevier, New York, 1973.
41. ICI Ltd. Tech. Bulletin on Saffil Fibers, 1975.
42. Allcock, H. R., *Chem. Brit.* **10**, 118 (1974).
43. Allcock, H. R., *Chem. Rev.* **72**, 315 (1972).
44. *C & E News* p. 18, May 26, 1975.
45. Willertz, L. E., and Shewman, P. G., *Metall. Trans.* **1**, 2217 (1970).
46. Hirth, J. P., and Pound, G. M., *Progr. Mater. Sci.* **11,** 1 (1963).
47. Bauer, E., and Poppa, H., *Thin Solid Films* **12**, 167 (1972).
48. Schlatter, J. C., *Int. Conf. Sintering Related Phenomena, 4th, Univ. of Notre Dame, Notre Dame, Indiana* May 26–28, 1975.
49. Norris, L. F., and Parravano, G., *in* "Reactivity of Solids" (J. W. Mitchell, ed.), p. 149. Wiley (Interscience), New York, 1969.
50. Baird, T., Paal, Z., and Thomson, S. J., *J. Chem. Soc.* **1**, 50 (1973).
51. Stonehart, P., and Zucks, P. A., *Electrochim. Acta* **17**, 2333 (1972).
52. Wynblatt, P., and Gjostein, N. A., *Progr. Solid State Chem.* **9,** 21 (1974).
53. Wanke, S. E., and Flynn, P. C., *Catal. Rev.-Sci. Eng.* **12**, 93 (1975).
54. Flynn, P. C., Wanke, S. E., and Turner, P. S., *J. Catal.* **33**, 233 (1974).
55. Ruckenstein, E., and Pulvermacher, B., *AIChE J.* **19**, 356 (1973).
56. Ruckenstein, E., and Pulvermacher, B., *J. Catal.* **29**, 224 (1973).
57. Flynn, P. C., and Wanke, S. E., *J. Catal.* **34**, 390 (1974).
58. Vannice, M. A., Benson, J. E., and Boudart, M., *J. Catal.* **16**, 348 (1970).
59. Hassan, S. A., *J. Appl. Chem. Biotechnol.* **24**, 497 (1974).
60. Emelianova, G. I., and Hassan, S. A., *Proc. Int. Congr. Catal., 4th, Moscow,* p. 1329 1968.
61. Maat, H. J., and Moscou, *Proc. Int. Congr. Catal. 3rd, Amsterdam,* p. 1277, 1964.
62. Hermann, R. A., Adler, S. F., Goldstein, M. S., and Debaun, R. M., *J. Phys. Chem.* **65**, 2189 (1961).
63. Kirklin, P. W., and Whyte, T. E., Jr., Am. Chem. Soc., Div. Pet. Chem. prepr., 163rd Nat. ACS Meeting, Boston, Massachusetts, p. C32, 1972.
64. Somorjai, G. A., Powell, R. E., Montgomery, P. W., and Jura, G., *in* "Small Angle X-ray Scattering" (H. Brumberger, ed.), p. 449. Gordon and Breach, New York, 1967.
65. Hughes, T. R., Houston, R. J., and Sieg, R. P., *Ind. Eng. Chem. Process Des. Dev.* **1**, 96 (1962).

66. Mills, G. A., Weller, S., and Cornelius, *Proc. Int. Congr. Catal., 2nd, Paris* p. 2295, 1960.
67. Furuoya, I., and Shirasaki, T., *Kogyo Kagaku Zasshi* **72**, 1223 (1969).
68. Furuoya, I., Shirasaki, T., Echigoya, I., and Morikawa, K., *Kogyo Kagaku Zasshi* **72**, 1431 (1969).
69. Baker, R. T. K., Thomas, C., and Thomas, R. B., *J. Catal.* **38**, 510 (1975).
70. Thomas, C. L., "Catalytic Processes and Proven Catalysts," pp. 54–63. Academic Press, New York, 1970.

Chapter 11

Surface Science

Work done over the last decade in surface science has influenced catalysis research by two broad developments: the evolution of models for surfaces, chemisorption, and other surface phenomena, and the progress in spectroscopic techniques for characterizing catalytic surfaces. This chapter will give a brief overview of the experimental and theoretical studies dealing with surfaces and a more detailed review on instrumental techniques.

11.1 CONCEPTS RELEVANT TO CATALYSIS

Most of the recent concepts in surface science applicable to catalysis can be conveniently divided into three areas: catalytic properties of surface overlayers; structural effects on surface reactions; and theoretical aspects of surface composition and of small clusters.

The influence of surface overlayers on catalysis has been known for some time, but only recently were attempts made to define the chemical and physical properties of these layers. An example of such an effort is the work of Madix and co-workers [1–9] concerned with the decomposition of formic and acetic acids on well-defined nickel and carbided nickel surfaces. Using flash desorption and Auger electron spectroscopy (AES) these authors have shown that surface carbide is selective for H_2 and CO_2 formation while a clean nickel surface gives predominantly H_2O and CO. The selectivity for CO_2 correlates well with the inability of the surface to adsorb CO_2. The results further suggest that the decomposition of formic acid to CO and the adsorption of CO_2 involve similar binding sites. The presence of ordered surface carbide significantly reduces the binding energy of CO, H_2O, CO_2, and H_2.

In more recent work, Madix and his co-workers have found that small amounts of oxygen tend to modify the adsorption on, and the catalytic properties of, the surface nickel atoms. It is thought that oxygen penetrates the nickel surface to the second layer below the surface [9]. The oxygen atoms then act as ligands which alter the electronic properties of the surface nickel. Excess oxygen eventually covers the surface and poisons it for further reaction.

In trying to understand the influence of the geometric or structural factor on catalysis, many researchers have studied the effect of the crystal orientation of the catalyst surface on the turnover number, the number of molecules reacting per second per surface site. Thus, the concept of structure-sensitive and structure-insensitive reactions which has been treated in Section 3.3 has evolved [10]. In the area of surface science, Somorjai and co-workers have been particularly active in this respect. In a study of the structure-sensitive nature of the dehydrocyclization of n-heptane to toluene [11,12] they used single crystals prepared to show flat surfaces or surfaces exhibiting steps and kinks, and found that, in the presence of hydrogen, the rate of this reaction is higher on stepped (111) surfaces than on flat (110) surfaces. Carbon–carbon bonds were found to break almost exclusively at the ledges which separate the terraces on the crystal. Hydrogen–hydrogen bonds and carbon–hydrogen bonds were also believed to undergo scission at these same sites, as well as at the steps. It was concluded that bonds do not break as readily on smooth planes than at steps and kink sites. Moreover, a degree of order of carbon overlayer which forms on these surfaces was found to affect some surface reactions. Cyclohexane hydrogenolysis occurs regardless of the structure of the overlayer, but other reactions such as the conversion of cyclohexene to benzene and of *n*-heptane to toluene require a high degree of ordering in the overlayer [13]. The structure of this overlayer can also influence selectivity. Thus, in the dehydrogenation of cyclohexane the overlayer is initially ordered, and the dominant product is benzene. As the reaction proceeds, the overlayer becomes disordered, and cyclohexene becomes the main product. Finally, it was found that the difference in the rates of bond breaking between planes and steps has an effect on various catalytic reactions. One example is the rate of dehydrogenation of cyclohexane to benzene which is faster by an order of magnitude on stepped surfaces than on the smooth planes [13].

Kahn *et al.* [14] studied the hydrogenation of cyclopropane over platinum single crystals at atmospheric pressure. The turnover number was similar to that measured for highly dispersed supported platinum catalysts. This work is significant because it showed for the first time that specific reaction rates were similar for single crystals and highly dispersed catalysts at comparable conditions. It also confirmed that the hydrogenation of cyclopropane is a structure-insensitive or facile reaction.

Another contribution from surface science of importance in catalyst poisoning, sulfiding, and activity maintenance is the concept of corrosive chemisorption. An example of this phenomenon is given by the work of Domange and Oudar [15] who found that the adsorption of sulfur on copper single crystals caused a reconstruction of the outermost layer of copper atoms into a two-dimensional sulfide layer. Low-energy electron diffraction showed that at saturation a monolayer of sulfur contains the same number of sulfur atoms on each of the three low index planes. Thus, the idea of specific sites becomes more diffuse since the same reconstructed surface layer is formed regardless of the original crystallographic orientation of copper.

Other metals behave differently. Platinum and nickel appear to show a specific interaction of certain crystal planes with sulfur, thereby altering selectivity as well as activity (see Section 6.2). It has been suggested [16] that this is due to a reconstruction of the surface of the catalyst as a consequence of the difference in surface energy of the various low-index planes of the metal. It is thought that the adsorption of small amounts of H_2S modifies the surface energy balance and leads to a new equilibrium distribution of surface planes with different catalytic activity. Such an explanation has important consequences for the types of reactions that are affected by poisoning or reconstruction by H_2S. Thus, a structure-sensitive reaction such as hydrogenolysis and isomerization would be affected to a much greater degree by this type of poisoning than a structure-insensitive reaction such as hydrogenation.

The studies in surface science concerning surface overlayers, structural effects, or surface reconstruction have led to a better understanding of some of the factors which can be important in selecting reaction paths of catalytic reactions. It is possible that continued efforts along these lines will give the catalytic scientist guidance for the modification of catalytic selectivity and for improvement of catalyst activity maintenance.

Two important contributions of theoretical surface science to catalysis are the prediction of surface composition of alloys and the theory of small clusters.

Because of the importance of alloy catalysis, a number of thermodynamic studies have been concerned with predicting surface and bulk composition for dispersed bimetallic systems. The most notable are those by Ollis [17], Hoffman [18], Williams [19], and Sachtler and co-workers [20] which predict changes in the composition of a surface as a function of temperature, chemical environment, and crystallite size. Many of these studies are based on solution theory, and conclude that the component with the lowest surface free energy will predominate at the surface. According to these models, the difference between bulk and surface compositions decreases exponentially with increasing temperature. In the case of NiCu alloys, Cu having the lower surface energy will be present on the surface in a higher concentration than in the bulk.

The surface composition of an alloy catalyst during operation can also be influenced by its reactive environment. Williams and Boudart [21] found that the chemisorption of O_2 on Ni–Au alloys increased the surface concentration of Ni. Similar results were obtained by Bartholomew and Boudart [22] for the surface concentration of Fe in the chemisorption of O_2 on Pt–Fe alloys and also by Sachtler and co-workers [23,24] for the surface concentration of Pd and Pt in the chemisorption of CO on Pd–Ag and Pt–Ag alloys. The latter authors observed no effect for the Pt–Ru system. These studies suggest that alloy characterization data such as selective chemisorption measurements of surface composition at ambient conditions should be used with caution for interpreting catalytic results.

The theory of small atomic clusters has been stimulated by an intense interest in the physics of small particles, primarily because of the usefulness of these materials in catalysis. Johnson [25] has applied the Xα scattered wave method to metal clusters, a quantum mechanical procedure originally developed for treating polyatomic molecules and solids. It allows the accurate calculation of the ground and excited electronic states of polyatomic clusters as functions of cluster size, composition, geometry, and environment. Its primary advantage is that it requires only moderate amounts of computer time compared to other approximation procedures. The application of this computational procedure to chemisorption on transition metal surfaces and to metal aggregates has recently been described [25], but so far insufficient work has been done to fully assess its potential.

Baetzold and co-workers [26–28] have also developed quantum mechanical computational procedures for determining the electronic and atomic structure of metal clusters. They have found in the case of Cu–Ni and Pd–Ag that, in contrast to expectations based on rigid band model calculations, d holes exist on the Group VIII metal atoms for compositions up to and greater than 60% Ib metal [26]. In this same area, work by Wertheim [29] and Spicer [30] has shown that the Anderson model or virtual bound state treatment for bimetallic systems is a better approximation than that provided by rigid band theory. In this model a bound d state is associated with each transition metal atom. The width of the band associated with this state depends upon alloy composition. For example, for Ni metal one starts with a broad d band which has 0.6 holes above the Fermi level. As copper is added to form a NiCu alloy, the band narrows, reduces in intensity, and the Fermi level effectively moves to the edge of the d band. It reaches the edge of the band at about 60% Cu which is precisely where ferromagnetism disappears. This model is in contradistinction to the rigid band model which pictures the addition of Cu as donating electrons to the empty d band with subsequent pairing and loss of magnetism.

TABLE 11-1

TECHNIQUES FOR MATERIALS ANALYSIS[a]

Input energy source	Observation	Name of technique	Information				Applicability				Interpretation	R and D in catalysis needed
			Surface	Bulk	Structure	Bonding	Catalytic reactions	Catalytic materials	Degree of generality	Maturity		
γ Rays	Absorption	Mössbauer effect	No[b]	Yes	Yes	Yes	Sometimes	Yes	Restricted	Developed	Simple to complex	Yes
X rays	Absorption	EXAFS	No[b]	Yes	Yes	Yes	No	Yes	Broad	New	Complex	Yes
	Fluorescence	—	No[b]	Yes	No	Yes	Yes	Yes	Broad	New	Simple	Yes
	Diffraction	—	No	Yes	Yes	No	Questionable	Yes	Broad	Highly developed	Routine	No
	Small-angle scattering	SAXS	No	Yes	Yes	No	Questionable	Yes	Broad	New	Complex	Yes
	Electrons	XPS/ESCA	Yes	Yes	No	Yes	Questionable	Yes	Broad	New	Complex	Yes
Ultraviolet radiation	Electrons	UPS	Yes	Yes	No	Yes	Possibly	Yes	Broad	New	Under development	Yes
Visible radiation	Reflection	—	No[b]	Yes	No	Yes	Yes	Yes	Broad	Undeveloped	Complex	Yes
	Reflection (polarized light)	—	No[b]	Yes	No	Possibly	Possibly	Possibly	Unknown	Undeveloped	Complex	Possibly
Infrared radiation	Attenuated total reflectance	ATR	Yes	No	No	Yes	Yes	No	Restricted	New	Simple to complex	Possibly
	Absorption	—	No[b]	Yes	No	Yes	Yes	Yes	Broad	Highly developed	Simple	No
	Absorption/ emission (Fourier transform)	FTS	Yes	Yes	No	Yes	Yes	Yes	Broad	New	Simple	Yes
	Laser Raman scattering	LRS	No[b]	Yes	No	Yes	Possibly	Possibly	Broad	New	Simple to complex	Yes
Microwaves	Absorption	NMR (wide band)	No[b]	Yes	No	Yes	Possibly	Sometimes	Restricted	Developed	Simple to complex	Yes
	Absorption	NMR (high resolution)	No[b]	No	No	Yes	Possibly	Possibly	Restricted	New	Under development	Yes
Radiowaves	Absorption	EPR	No[b]	Yes	Sometimes	Yes	Possibly	Yes	Restricted	Developed	Complex	Possibly
Magnetic fields	Magnetization	Magnetic susceptibility	No[b]	Yes	Sometimes	Yes	Possibly	Yes	Restricted	Highly developed	Moderately complex	No

TABLE 11-1 *(continued)*

Input energy source	Observation	Name of technique	Information				Applicability				Interpretation	R and D in catalysis needed
			Surface	Bulk	Structure	Bonding	Catalytic reactions	Catalytic materials	Degree of generality	Maturity		
Moderate electric fields	Electric potential	Surface potential	Yes	No	No	Sometimes	No	Yes	Restricted	Developed	Simple	No
	Resistivity	Electrical conductivity	No[b]	Yes	No	No	Possibly	Yes	Restricted	Developed	Simple	No
	Electric potential	Hall effect	No[b]	Yes	No	No	Possibly	Yes	Broad	Undeveloped	Simple	No
High electric fields	Field emitted ions	FEIM, FIM	Yes	No	Yes	Yes	Doubtful	No	Restricted	Moderately developed	Complex	Yes
	Field emitted electrons	FEEM, FEM	Yes	No	Yes	Yes	Doubtful	No	Restricted	Developed	Complex	No
	Field emitted electrons	FEED	Yes	No	No	Yes	No	No	Restricted	New	Under development	No
High-energy electrons	Diffraction	HEED	No	Yes	Yes	No	No	No	Restricted	Developed	Under development	No
	Energy loss	ELS	Yes	No	No	Yes	No	Possibly	Broad	New	Under development	Yes
	X-rays	APS	Yes	Yes	No	Yes	No	Possibly	Broad	New	Under development	Possibly
Low-energy electrons	Diffraction	LEED	Yes	No	Yes	No	No	No	Restricted	Developed	Under development	No
	Scattering	ILEED LEES	Yes	No	No	Yes	No	No	Unknown	New	Under development	No
Ions	Electrons	INS	Yes	No	No	Yes	No	No	Unknown	New	Complex	No
	Ions	EIS	Yes	No	No	No	No	Yes	Broad	New	Simple	Yes
	Ions	SIMS	Yes	No	No	No	No	Possibly	Unknown	New	Simple	No
Heat at programmed rate	Gas desorption	TPD	Yes	No	No	Yes	No	Yes	Broad	New	Simple	Yes
	Reaction produces desorption	TPK	Yes	No	No	Yes	Yes	No	Broad	Undeveloped	Simple	No
None	Heat	Micro-calorimetry	No[b]	Yes	No	Yes	Yes	Yes	Broad	Developed	Simple	No

[a] From **Boudart**, courtesy of the Electric Power Research Institute.

[b] Surface information can be obtained from these techniques for well-dispersed catalysts or when the interaction of the input energy occurs with species that are primarily on the surface.

TABLE 11-2
DEFINITION OF ACRONYMS USED IN TABLE 11-1[a]

Acronym	Definition
EXAFS	Extended x-ray absorption fine structure
SAXS	Small-angle x-ray scattering
XPS/ESCA	X-ray photoelectron spectroscopy
UPS	Ultraviolet photoelectron spectroscopy
ATR	Attenuated total reflectance
FTS	Fourier transform spectroscopy
LRS	Laser Raman spectroscopy
NMR	Nuclear magnetic resonance
EPR	Electron paramagnetic resonance
FEIM/FIM	Field emission ion microscopy
FEEM/FEM	Field emission electron microscopy
FEED	Field electron energy distribution
HEED	High-energy electron diffraction
ELS	Energy loss spectroscopy
AES	Auger electron spectroscopy
APS	Appearance potential spectroscopy
LEED	Low-energy electron diffraction
ILEED	Inelastic low-energy electron diffraction
LEES	Low-energy electron scattering
INS	Ion neutralization spectroscopy
EIS/ISS	Elastic ion scattering spectroscopy
SIMS	Secondary ion mass spectroscopy
TPD (TDS)	Temperature programmed desorption (thermal desorption spectroscopy)
TPK	Temperature programmed kinetics

[a] From Boudart, courtesy of the Electric Power Research Institute.

11.2 PHYSICAL TECHNIQUES

Numerous physical methods developed in surface science having actual and potential applications in the study of catalysts and catalytic reactions are summarized in Tables 11-1 and 11-2 [31].

The techniques which have had the most significant impact in catalysis fall in two categories, electron and ion spectroscopies. Discussion of the most useful spectroscopies follow.

11.2.1 Electron Spectroscopies

In *x-ray photoelectron spectroscopy (XPS)* [32] an x ray of known energy $h\nu$ is absorbed by a sample, ejecting electrons with kinetic energies given by $E_K = h\nu - E_B - \phi$, where E_B is the binding energy of the electrons with respect to the

Fermi level, and ϕ is the work function of the sample. Since the sample is in contact with the spectrometer, ϕ is also the work function of the spectrometer. Therefore, the technique provides direct measurement of E_B, the energy level of the core electrons, yielding atomic identification. The chemical shifts frequently observed for these energies can be used to obtain information on the distribution of the valence electrons in a solid [32–35] and to obtain a description of bonding in terms of the partial charges on the atoms.

In *ultraviolet photoelectron spectroscopy (UPS)* [36,37] photon excitation energies are not high enough to eject core electrons, and therefore only valence electrons are studied. The advantages of using this technique for valence electron studies are higher resolution and greater signal intensity compared to XPS. The increased resolution occurs because vacuum uv light source linewidths are only a few million electron volts, whereas soft x-ray linewidths (XPS) are of the order of 1 eV. The reason for increased intensity is primarily related to the high photon fluxes and larger ionization cross section for the valence electrons with uv radiation. A drawback of high photon fluxes is the possible damage of the surface under study by photodesorption. As with XPS, chemical shifts are also observed.

Auger electron spectroscopy (AES) [38,39] involves an energy analysis of secondary electrons (Auger electrons) which are ejected when a core hole relaxes. The initial hole can be produced by several excitation sources including x rays or electrons. Auger transitions do not involve valence electrons and readily identify the atom. Chemical shift data are more difficult to interpret because three energy levels are involved. Thus, AES has been used primarily for atomic identification or surface analysis [38,39].

These three spectroscopies represent surface analysis techniques because most of the information obtained comes from the first several atomic layers. One of the limitations of XPS, UPS, and AES is that the sampling depth can vary with excitation energy and the form of the sample (e.g., powder, flat crystal). A recent review [40] shows that the sampling depth is in general about a factor of two smaller for AES than for XPS (primarily because of lower kinetic energies of escaping electrons) and that UPS escape depths may vary considerably. Typical ranges are 2–24 Å for AES, 7–40 Å for XPS, and 3–50 Å for UPS. This variation in sampling depth causes difficulties in quantitative measurements. AES has been used more often and more successfully for this purpose than the other two techniques. By choosing the proper energy line, XPS can be used for quantitative measurements. UPS, on the other hand, is not as adaptable for this purpose. This is so because the spectra do not lend themselves to direct identification of elements as they usually consist of broad peaks superimposed on a scattered electron background [40].

A major problem with these techniques has been the effects of sample charging which can be significant for insulators used as catalyst supports (e.g., Al_2O_3). In investigating the chemical state of surface atoms (chemical shifts), this phenomenon can have a marked effect on interpretation of the spectra. Many workers have tried to dope insulator samples with various standards such as gold, and then to compensate for shifts with a known spectrum without obtaining entirely reproducible data. Most of the surface science work in this area has been done with metals, for which charging has not been a major consideration.

Because of low pressure (perhaps ultimately up to 10^{-2} torr) requirements, it is unlikely that industrial catalytic reactions can be effectively studied with any of the above three electron spectroscopies. The major use of these techniques appears to be in characterizing a surface before and after reaction, or in the study of chemisorption. AES has been applied primarily for surface analysis, while UPS and XPS have been used to study the chemical state of surface atoms or the nature of the interaction of chemisorbed species. The following examples are representative of this work.

Using AES, Szalkowski and Somorjai [41] studied the surface composition of vanadium oxides on vanadium single crystals. The observed chemical shifts were 0.6 eV per unit nominal oxidation number for inner-shell Auger transitions. Using chemical shifts, and the oxygen to vanadium Auger peak intensity ratios for the various vanadium oxides ($VO_{0.92}$, V_2O_3, V_2O_5, and VO_2), it was possible to determine the surface composition of the different oxides formed in the oxidation of vanadium metal.

More recently, Takeuchi *et al.* [42] studied the dependence of the activity of nickel sulfide catalysts for the selective hydrogenation of alkynes on their surface composition estimated by AES. These workers found that a nickel sulfide catalyst having an atomic ratio of S/Ni = 0.6 had good activity for selective hydrogenation of acetylene to ethylene, while a nickel sulfide catalyst with S/Ni = 0.9 had no measureable activity. According to a model developed, the nickel atom is surrounded by four sulfur atoms in Ni_3S_2 and by six sulfur atoms in NiS. Therefore, it was reasoned that nickel atoms on the surface of Ni_3S_2, but not on that of NiS, have the appropriate coordinative unsaturation for the chemisorption of acetylene and that partial hydrogenation is controlled by the reversible formation of such coordinatively unsaturated nickel atoms. These results are in agreement with work by Kirkpatrick [43] who showed that NiS is inactive for the partial hydrogenation of polyenes to corresponding monoolefins, but becomes active with time due to the reduction of the sulfide to Ni_3S_2.

Using AES, Williams and Baron [44] followed the poisoning of platinum and palladium oxidation catalysts by lead, phosphorus, and sulfur contaminants in automotive exhaust. They found that the noble metal surface is first saturated with lead without any deposition of sulfur or phosphorus. The initial accumulation of Pb on Pd is different from that for Pt. At low exposures, Pb does not accumulate on the Pd surface but diffuses as PbO from the surface into the bulk of the Pd. The latter is in the oxide form during operation, the diffusion resulting from the solubility of PbO in PdO. On Pt, however, this cannot occur because

PtO is not formed at these conditions. Continued exposure to S, P, and Pb leads to the deposition of $PbSO_4$ and $Na_4P_2O_7$.

It has also been found by means of AES that a gaseous environment can change the surface composition of an alloy. Thus Williams and Boudart [21] observed that the surface composition of Ni–Au alloys over a wide range of compositions was predominantly Au. However, on exposure of these alloys to O_2, no surface Au could be detected, as oxygen drew nickel to the surface.

The use of UPS and XPS has been primarily directed at the nature of the surface orbitals and the molecular orbitals of chemisorbed molecules [45–47]. For example, for heteronuclear diatomic molecules, characteristic photoemission from either atom in the adsorbed molecule may be studied [37]. Oxygen lines have been used to identify the two types of adsorbed CO species on tungsten, the α and β forms. The O (1s) chemical shifts from the XPS technique have been used to follow the desorption of α-CO species from the CO monolayer. Similar work at the National Bureau of Standards [37,46] with N_2, NO, O_2, H_2CO, and CO indicates that the 1s binding energies of adsorbate atoms decrease as the strength of adsorption increases in qualitative agreement with physical models [48].

Recent UPS studies of C_2H_2, C_6H_6, and CH_3OH have been reported [48,49] which are thought to show the nature of surface and bond formation and dissociation. For C_6H_6 adsorbed in Ni (111), for example, both σ and π bonding of the hydrocarbon have been observed in the chemisorbed layer. With CH_3OH on a W(100) surface, UPS indicates that dissociation occurs at low coverages to yield adsorbed CO and H atoms. As coverage increases, more complex species, possibly CH_3O, are observed [49–51]. The advantage of UPS for these chemisorption studies is that the adsorbate is not as much affected by the excitation beam as it is with XPS and AES because photodissociation and photodesorption cross sections (probabilities) are very small for most adsorbed species [51].

Using UPS, Spicer *et al.* [52] looked at the change in surface electronic structure (valence band) for MoS_2, Cu, Pt, and Si upon physisorption and chemisorption of O_2, CO, H_2, and N_2. Physisorption caused only negligible electronic changes, but on chemisorption a significant drop in surface emission occurred which was attributed to hybridization of the orbitals of the surface and of the adsorbed gas. It was suggested that in the case of Pt this drop in emission indicates the metallic orbitals which participate in bonding. With Cu this hybridization is reported to produce new orbitals located within or above the d band. It would be of interest to apply recent theories for chemisorption, particularly that of Schrieffer [53,54], to these results.

UPS has also been used for probing the surface electronic structure of metals and alloys. For example, Helms *et al.* [55] studied the surface electronic structure, composition, and CO binding energy for Cu–Ni alloys. Individual peaks in the surface electronic structure were assigned to Ni and Cu. This is in contrast to what would be expected from rigid band theory. The binding energy of CO on the alloy was found to vary in a quasi-continuous manner with surface composition from that on Ni to that on Cu. It was concluded that the binding energy is determined by the local configuration of the adsorption site. This is in contrast with a model depending upon the average surface electronic structure.

The XPS studies of Ratnasamy *et al*. [56–59] on commercial cobalt molybdate catalysts indicate that there are at least three states for cobalt on the surface, Co^{+2} in a $CoAl_2O_4$ environment, Co^{+2} in a CoO phase, and Co^{+2} in coordination with strongly bound water molecules. Although the corresponding broad Mo peaks cannot be resolved, from reduction, sulfidation, and hydrodesulfurization studies [56–59] it is concluded that molybdenum is present in at least three phases: phase A in which chemical interaction with the support is so strong that the Mo ions cannot be reduced in hydrogen even at high temperatures; phase B which has minimal interaction with the support and is easily reduced to a suboxide or metal; and phase C which is partially reduced in the presence of cobalt metal.

11.2.2 Ion Spectroscopies

Only two ion spectroscopic techniques have received major attention in connection with catalyst problems. They are secondary ion mass spectrometry (SIMS) and ion scattering spectroscopy (ISS). Both use ion bombardment of the surface in order to obtain information on it.

In SIMS [60,61] the secondary ions emitted by the surface are identified by a mass spectrometer. ISS depends on the measurement of the energy of the ions which are scattered elastically by collision with surface [62–65].

For these ion techniques the incident beam consists of noble gas ions with energies ranging from 1 to 50 keV. In the low-energy region of this range, the probability for neutralization is very high [62] (greater than 99.99% for He^+ ions), but a sufficient number of ions are scattered to allow detection. In fact, the high neutralization probability assures that those ions which are detected in ISS result from single scattering events. For these particles a simple two-body collision model adequately describes the mechanics of the system. It leads to a relationship among incoming and outgoing energies, the scattering angle, and the masses of the incoming ion and that of the surface atoms. It is therefore possible to identify the surface atoms by mass. By the principle of conversion of energy and momentum for a binary elastic collision between an energetic noble gas ion of mass M_1 and a surface atom of mass M_2, the following relationship is easily derived:

$$M_2 = M_1\,[(1 + (E_1/E_0))/(1 - (E_1/E_0))]$$

In this derivation, the scattering angle is taken as 90° and E_0 and E_1 are the energies of the ion beam before and after scattering, respectively.

When the energy of the incoming ions is increased, the scattering event becomes more complex and multiple scattering occurs. Since this scattering is a function of the relative position of the surface atoms, analysis of the energy spectrum can, in principle, provide information on the short-range structure of these surfaces [62].

The primary complication of the techniques using high-energy ions is the possibility of disturbing the surface by the impact of ions. Sputtering of surface atoms may not proceed uniformly, thereby leading to surface concentration changes. The ion techniques are not significantly affected by the conductivity of the solid, and therefore the sample charging encountered for electron spectroscopies is not a problem. Also, ion techniques are sensitive to the presence of hydrogen, which cannot be detected by electron spectroscopic techniques. Hydrogen detection could be important in the determination of the composition of C–H residues on catalyst surfaces. Work in this area is under way at the University of California at Berkeley by Somorjai and co-workers [66].

A number of detailed papers describing SIMS, ISS, and other ion spectroscopies can be found in Ref. [67].

The application of SIMS and ISS to catalytic materials has not been extensive, more work having been done with thin films and electronic devices. However, the following two studies illustrate the potential of these techniques.

Shelef *et al*. [68] examined the surface composition of the spinel structures $CuAl_2O_4$, $CoAl_2O_4$, and $NiAl_2O_4$ and of other oxides using ISS. The results showed no cobalt on the surface of $CoAl_2O_4$, but showed considerable amounts of transition metals on the surface of $CuAl_2O_4$ and $NiAl_2O_4$, and correlated well with NO surface area measurements (see Section 4.3.2), indicating negligible adsorption for $CoAl_2O_4$, but substantial adsorption for the other two compounds. These results are important because they explain the reason for the loss of the oxidation activity of Co_3O_4 when supported on Al_2O_3. This combination forms a spinel, the surface of which is completely covered with oxygen, and has therefore low activity. The compounds CuO and NiO also form spinels when supported on Al_2O_3, but retain some of their oxidation activity because the active constituents, presumably Cu^{+2} and Ni^{+2} cations, are still accessible to adsorbing gases, and are not covered by oxygen anions. These facts are in qualitative agreement with various solid-state models [69] for these three spinels which consider $CoAl_2O_4$ a "normal" spinel with Co^{+2} ions in tetrahedral positions, while Ni^{+2} ions have an octahedral site preference. Cations in the tetrahedral sites are electrically more unstaurated if one anion is missing on the surface from their coordination sphere. These cations will tend to diffuse below the surface and be shielded by oxygen anions more than cations in the octahedral sites. Similar but more complex arguments involving distorted mixed structures can be made for the Cu spinels [69,70].

Other work by Wheeler and Bettman [71] used both ISS and SIMS to study the poisoning of Co_3O_4 by Al_2O_3 and MgO. The specific oxidation activity of

Al_2O_3-doped Co_3O_4 for hydrocarbon oxidation was found to decrease by a factor of about 4–5, while for the MgO-doped Co_3O_4 this factor was about 8, indicating that the surface of the Al_2O_3-doped sample was enriched in Al relative to the bulk and that of the MgO-doped sample was depleted of Co. Similar results were obtained for $MgAl_2O_4$-supported Co_3O_4 in which the poisoning effect was ascribed to the leaching of MgO and Al_2O_3 from the support by the strongly acidic cobalt nitrate solution used in catalyst preparation. This was confirmed by using catalysts prepared with a neutralized cobalt solution.

These examples show clearly that SIMS and ISS have some application in catalyst characterization, especially where profiles of composition are required. However, there are a number of limitations such as variations in sputtering rate with environment or species. More work needs to be done to assess these physical techniques.

11.3 SUMMARY AND CONCLUSIONS

Surface science has led to improved methods for catalyst characterization by electron spectroscopies (mainly AES, XPS, and UPS) and ion spectroscopies (SIMS and ISS). All of these techniques have limitations, but these are being gradually overcome as shown by the following examples.

Wagner [72–75] has reported on an improved XPS using core Auger lines which may obviate the severe problem of sample charging, and thereby give universally reproducible data. Stern and co-workers [76–78] have developed a novel x-ray technique called extended x-ray absorption fine structure spectroscopy (EXAFS) which may be able to identify surface atoms and indicate their local structural and electronic environment. This method has already been applied to some catalysts such as $CuCrO_3$ [79], Au/MgO [80], and Pt/Al_2O_3 [80,81] and may be readily applied to catalysts during reaction [82,83]. There are some researchers who question the applicability of the techniques of surface science to catalysts because they were developed for well-defined, clean surfaces. Although caution is most certainly warranted, this is perhaps an extreme viewpoint. According to another view [84] catalytic scientists adopting the methods of surface science will adapt them to elucidate catalytic phenomena and the nature of catalytic materials.

REFERENCES

1. Madix, R. J., Falconer, J., and McCarty, J., *J. Catal.* **31**, 316 (1973).
2. McCarty, J., Falconer, J., and Madix, R. J., *Surface Sci.* **42**, 329 (1974).
3. McCarty, J., Falconer, J., and Madix, R. J., *J. Vac. Sci. Technol.* **11**, 266 (1974).
4. McCarty, J. G., and Madix, R. J., *J. Catal.* **38**, 402 (1975).
5. Falconer, J. L., and Madix, R. J., *Surface Sci.* **48**, 393 (1975).

6. McCarty, J., Falconer, J., and Madix, R. J., *J. Catal.* **30**, 235 (1973).
7. Madix, R. J., and Falconer, J. L., *Surface Sci.* **51**, 546 (1975).
8. Madix, R. J., Falconer, J. L., and Suszko, A. M., *Surface Sci.* **54**, 6 (1976).
9. Madix, R. J., private communication, 1976.
10. Boudart, M., *Adv. Catal.* **30**, 453 (1969).
11. Somorjai, G. A., Joyner, R. W., and Lang, B., *Proc. R. Soc. London* **331**, 335 (1972).
12. Somorjai, G. A., *Catal. Rev.* **7**, 87 (1972).
13. Somorjai, G. A., *C & E News*, p. 23, December 8, 1975.
14. Kahn, D. R., Peterson, E. E., and Somorjai, G. A., *J. Catal.* **34**, 294 (1974).
15. Domange, J. L., and Oudar, J., *Surface Sci.* **11**, 124 (1968).
16. Somorjai, G. A., *J. Catal.* **27**, 453 (1972).
17. Ollis, D. F., *J. Catal.* **23**, 131 (1971).
18. Hoffman, D. W., *J. Catal.* **27**, 374 (1972).
19. Williams, F. L., *Diss. Abstr. Int. B* **33**, 5807 (1973).
20. Sachtler, W. M. H., and Jongepier, R., *J. Catal.* **4**, 665 (1965).
21. Williams, F., and Boudart, M., *J. Catal.* **30**, 438 (1973).
22. Bartholomew, C., Surface Composition and Chemistry of Supported Platinum–Iron Alloy Catalysts, PhD. Thesis, Stanford Univ., 1972.
23. Bouwman, R., and Sachtler, W. M. H., *J. Catal.* **26**, 63 (1972).
24. Bouwman, R., and Sachtler, W. M. H., *J. Catal.* **19**, 127 (1970).
25. Johnson, K. H., and Messmer, R. P., *J. Vac. Sci. Technol.* **11**, 236 (1974).
26. Baetzold, R. C., and Mack, R. E., *J. Chem. Phys.* **62**, 1513 (1975).
27. Baetzold, R. C., *J. Chem. Phys.* **55**, 4363 (1971).
28. Baetzold, R. C., *Surface Sci.* **36**, 123 (1972).
29. Hufner, S., Wertheim, G. K., Cohen, R. L., and Wernick, J. H., *Phys. Rev. Lett.* **28**, 488 (1972).
30. Seib, D. H., and Spicer, W. E., *Phys. Rev. B* **2**, 1676, 1694 (1970).
31. Boudart, M., Cusumano, J. A., and Levy, R. B., New Catalytic Materials for the Liquefaction of Coal, Rep. RP-415-1, Electric Power Research Institute, October 30, 1975.
32. Siegbahn, K. *et al.*, *Nova Acta Roy. Soc. Sci. Ups.* **20** (1967).
33. Frost, D. C., Herring, F. G., McDowell, C. A., and Woolsey, I. S., *Chem. Phys. Lett.* **13**, 391 (1972).
34. Snyder, L. C., *J. Chem. Phys.* **55**, 95 (1971).
35. Ellison, F. O., and Larcom, L. L., *Chem. Phys. Lett.* **13**, 399 (1972).
36. Turner, D. W., Baker, A. D., Baker, C., and Brundle, C. R., "Molecular Photoelectron Spectroscopy." Wiley, New York, 1970.
37. Yates, J. T., Jr., *C & E News*, p. 19, August 26, 1974.
38. Palmberg, P. W., *in* "Electron Spectroscopy" (D. A. Shirley, ed.), p. 835. North-Holland Publ., Amsterdam, 1972.
39. Chang, C. C., *Surface Sci.* **25**, 53 (1971).
40. Brundle, C. R., *J. Vac. Sci. Technol.* **11**, 212 (1974).
41. Szalkowski, F. J., and Somorjai, G. A., *J. Chem. Phys.* **56**, 6097 (1972).
42. Takeuchi, A., Tanaka, K., Toyoshima, I., and Miyahara, K., *J. Catal.* **40**, 94 (1975).
43. Kirkpatrick, W. J., *Adv. Catal.* **3**, 329 (1951).
44. Williams, F., and Baron, K., *J. Catal.* **40**, 108 (1975).
45. Shirley, D. A., *Chem. Phys. Lett.* **16**, 220 (1972).
46. Yates, J., Jr., Madey, T., and Erickson, N., *Surface Sci.* **43**, 257 (1974).
47. Atkinson, S., Burndel, C., and Roberts, M., *Chem. Phys. Lett.* **24**, 175 (1974).
48. Shirley, D. A., *Chem. Phys. Lett.* **16**, 220 (1972).
49. Eastman, D., and Demuth, J., *Phys. Rev. Lett.* **32**, 1123 (1974).

50. Plummer, E., Waclawski, B., and Vorburger, T., *Chem. Phys. Lett.* **28**, 510 (1974).
51. Eastman, D., and Nathan, M. I., *Physics Today* p. 44, April 1975.
52. Spicer, W. E., Yu, K. Y., Pianetta, P., Lindau, I., and Collins, D., *in* "Surface and Defect Properties of Solids" (J. M. Thompson and M. W. Roberts, eds.), Vol. 5. Univ. of Bradford, The Chemical Society, London, 1972.
53. Schrieffer, J. R., *J. Vac. Sci. Technol.* **9**, 561 (1971).
54. Paulson, R. H., and Schrieffer, J. R., *Surface Sci.* **48**, 329 (1975).
55. Helms, C. R., Yu, K. Y., and Spicer, W. E., *Solid State Commun.* **18,** 365 (1976).
56. Ratnasamy, P., *J. Catal.* **40**, 137 (1975).
57. Ratnasamy, P., Mehrotra, R., and Ramaswamy, A., *J. Catal.* **32**, 63 (1974).
58. Ratnasamy, P., Sharma, L., and Sharma, D., *J. Phys. Chem.* **78**, 2069 (1974).
59. Ratnasamy, P., Ramaswamy, A., Banerjec, K., Sharma, D., and Ray, N., *J. Catal.* **38**, 19 (1975).
60. Rubin, S., *Nucl. Instrum. Methods* **5**, 177 (1959).
61. Werner, H., *Surface Sci.* **47**, 301 (1975).
62. Brogersma, H. H., *J. Vac. Sci. Technol.* **11**, 231 (1974).
63. Goff, R. F., *J. Vac. Sci. Technol.* **10**, 355 (1973).
64. Taglauer, E., and Heiland, W., *Surface Sci.* **47**, 234 (1975).
65. Niehus, H., and Bauer, E., *Surface Sci.* **47**, 222 (1975).
66. Somorjai, G. A., private communication, 1976.
67. *Proc. Conf. Ion Beam Surface Layer Anal., Yorktown Heights, New York* (June 1973).
68. Shelef, M., Wheeler, M., and Yao, H., *Surface Sci.* **47**, 697 (1975).
69. Cimino, A., and Schiavello, M., *J. Catal.* **20**, 202 (1970).
70. Lo Jacono, M., Schiavello, M., and Cimino, A., *J. Phys. Chem.* **75**, 1044 (1971).
71. Wheeler, M., and Bettman, M., *J. Catal.* **40**, 124 (1975).
72. Wagner, C., *Anal. Chem.* **44**, 1050 (1972).
73. Wagner, C., and Biloen, P., *Surface Sci.* **35**, 82 (1973).
74. Wagner, C., *Discuss. Faraday Soc.* **60,** 291 (1975).
75. Wagner, C., *Anal. Chem.* **47**, 1201 (1975).
76. Stern, E., *Phys. Rev. B* **10**, 3027 (1974).
77. Lytle, F. W., Sayers, D., and Stern, E., *Phys. Rev. B* **11,** 4825 (1975).
78. Stern, E., Sayers, D., and Lytle, F., *Phys. Rev. B* **11**, 4836 (1975).
79. Lytle, F. W., Sayers, D., and Moore, E., Jr., *Appl. Phys. Lett.* **24,** 45 (1974).
80. Bassi, I. W., Lytle, F. W., and Parravano, G., *J. Catal.* **42,** 139 (1976).
81. Lytle, F. W., *J. Catal.* **43,** 376 (1976).
82. Sinfelt, J. H., private communication, 1976.
83. Shelef, M., Dalla Betta, R., and Boudart, M., private communication, 1976.
84. Boudart, M., *Chemtech* p. 748, December, 1974.

PART III

SPECIFIC COAL CONVERSION PROCESSES

Chapter 12

Petroleum Products and Refining

This chapter serves as an introduction to subsequent chapters concerned with the upgrading and refining of coal liquids produced by the various conversion processes.

The first section gives a summary of the physical and chemical characteristics of the fuels conventionally derived from petroleum which are now being considered as potential products from coal liquids. The second section describes briefly the processing techniques, catalytic and otherwise, of petroleum refining needed to produce the various fuels.

12.1 FUEL PRODUCTS

All of the conventional petroleum fuel products are potentially available from coal liquids. A brief description of these fuels follows.

12.1.1 Motor Gasoline

This fuel is a blend of paraffins, naphthenes, and aromatics. It contains a substantial fraction of branched paraffins, benzene, toluene, and xylenes (BTX), the typical high-octane components. Optimization of characteristics, such as volatility, engine starting, antivapor lock, prevention of gum and sludge formation, as well as antiknock, requires careful blending of these components and depends also on climatic conditions.

12.1.2 Jet Fuel

Jet fuel is a highly paraffinic kerosene fraction with an aromatic content limited to less than about 20 vol%. This gives better control of smoke emissions and optimum energy density. The latter property is important for aircraft range and is set at a minimum of 18,400 Btu/lb. The energy density increases with an increasing H/C ratio and is therefore favored by paraffins and naphthenes. Polycyclic aromatics increase smoking, lead to excessive carbon deposition in the engine, and are therefore kept below 3 vol%. Some monocyclic aromatic content is desirable to improve the rich mixture performance of the jet aircraft during takeoff. Olefins must be held below 5 vol% because they decrease thermal stability and form polymeric gums. Sulfur and mineral constituents present potential problems because they form deposits and facilitate corrosion in the hot section of the jet engine. Mercaptans are generally limited to a maximum of 0.003 wt%.

12.1.3 Gas Turbine Fuels

The specifications for gas turbine fuels demand a minimum of 11.3 wt% of hydrogen, less than 1% sulfur, and low mineral matter, particularly sodium and potassium (<1 ppm), lead (<1 ppm), vanadium (as low as 0.5 ppm), and calcium (<10 ppm). The nitrogen concentration has to be below environmental specifications for total NO_x emissions. Other physical characteristics of concern include specific gravity, viscosity, carbon residue (less than 1%), flash point, and pour point [1].

12.1.4 Diesel Fuel

The measure of ignition quality for diesel fuel is called the cetane number, which is equal to 100 for cetane (*n*-hexadecane, $C_{16}H_{34}$), zero for α-methylnaphthalene, and intermediate for their mixtures. The cetane numbers usually vary from 35 to 50, depending on use, paraffinic fuels having better ignition qualities than aromatic fuels.

Additional important characteristics are volatility, viscosity, specific gravity, sulfur content, storage stability, pour point, cloud point, and flash point. High-speed diesel engines require volatile fuels because high boiling fractions cannot be vaporized quickly enough for combustion. Incomplete combustion causes carbon deposition and increased engine wear and total emissions. Excessively high viscosity results in the fuel spray (fuel injection) penetrating too far into the combustion chamber with subsequent wetting of cylinder walls and carbonization on the hot chamber surface. A low viscosity leads to engine wear since the fuel also serves as a lubricant. Both viscosity and volatility are controlled by the molecular weight of the paraffinic fractions.

Low sulfur levels are required for diesel engines to minimize emissions and to avoid corrosion and sludge deposits caused by sulfuric acid formed during combustion.

Typical uses of diesel fuel range from low-speed engines such as marine diesels to medium-speed railroad diesels as well as high-speed applications for buses, trucks, and tractors. The boiling range of the fuel varies from 325–550°F to 350–800°F, depending on the application. The sulfur constraint ranges from 1.2 wt% for low-speed use to 0.12 wt% for high-speed use. Nitrogen content varies with the local requirements for total NO_x emissions.

12.1.5 Heating Oils

The most frequently used heating oil is called #2 oil. It has an API gravity of 34, a viscosity (Saybolt Seconds Universal @ 100°F) of 35, and a boiling range of 325–645°F. Other specifications require a maximum carbon residue of 0.10 wt%, a pour point of −5°F, a flash point of 150°F, and a maximum sulfur content of 1 wt%. The oil should have proper viscosity to give optimum atomization and burning characteristics and minimum carbon deposition. The sulfur level is controlled for environmental reasons and to minimize burner corrosion.

12.1.6 Residual Fuels

Residual fuels are classified into several grades because of a variety of commercial applications, each having different requirements. The lightest residual fuel oil fraction, called #4 oil, requires no preheating. Grade #5 oil is used for oil burner installations provided with a preheater while the heaviest fraction, the #6 oil or Bunker C oil, needs special high-temperature preheating equipment.

Residual fuels usually contain a residuum or bottoms from atmospheric or vacuum distillation and a diluent. Diluents include coker gas oil, visbreaker gas oil, virgin gas oil, catalytic heating oil, and heavy catalytic cycle stock. The amount of diluent depends on the grade of the fuel oil and decreases with increasing grade number. Properties of primary importance for residual oils are viscosity and sulfur content, the latter being limited by local SO_x emission standards.

12.2 PETROLEUM REFINING

There are three types of modern refineries: hydroskimmer, conversion, and maximum conversion. The hydroskimmer refinery is the simplest of all and predominantly produces fractions by distillation from an atmospheric pipestill, namely gasoline (made by upgrading the naphtha fraction by catalytic reforming), jet fuel, diesel oil, heating oil, and fuel oil. This type of refinery is used primarily in Europe and other areas where fuel oil and heating oil are in high demand, and no significant incentive exists to maximize the gasoline fraction.

The conversion refinery is next in complexity and is extensively used in the United States to convert more than half of the crude oil into gasoline. The additional processing equipment required by the conversion refinery includes a vacuum pipestill to recover higher molecular weight fractions from the crude, a catalytic cracking unit for converting these fractions to more desirable products, and several hydrotreating units.

The maximum conversion refinery is similar to the conversion refinery but provides additional processing steps for the light and heavy ends to maximize middle distillate fractions. These steps include alkylation of paraffins, and hydrocracking and coking of heavy fractions.

Of the catalytic processes used, the most important ones for upgrading coal liquids are catalytic cracking, hydrocracking, naphtha reforming, and hydrofining. The latter process encompasses removal of sulfur, nitrogen, and oxygen compounds.

In catalytic cracking, a heavy feed, usually virgin gas oil, bp 600–1050°F, containing aromatics, naphthenes, and paraffins, is contacted with an acidic catalyst at 800–980°F and near atmospheric pressures in a fluidized bed which allows the withdrawal of the catalyst for oxidative regeneration. The objective is to produce gasoline, heating oil, and some diesel and jet fuel from heavy feedstocks with the formation of the minimum amount of light gases and of the optimum amount of coke. The coke is needed to provide heat on its oxidative removal from the catalyst during regeneration to balance the endothermicity of the cracking step. In addition to cracking, isomerization, alkylation, and dehydrogenation also occur.

Hydrocracking accomplishes the same function as catalytic cracking with higher molecular weight feedstocks and better control of product distribution. It is carried out at hydrogen pressures of a few thousand psig, using a bifunctional catalyst which contains both acid and hydrogenation sites.

Catalytic reforming converts a naphtha containing paraffins, naphthenes, and aromatics to high-octane gasoline over a bifunctional catalyst such as Pt/Al_2O_3 by maximizing the production of branched paraffins and single-ring aromatics. The process usually operates at 200–600 psig, 800–980°F, and at hydrogen flow rates of 5000–8000 SCF/bbl of naphtha.

Hydrofining operation invoves hydrodesulfurization (HDS) or hydrodenitrogenation (HDN) of the various streams in a refinery. It is carried out at pressures of up to 1500 psig at 600–800°F with hydrogen feed rates up to 5000 SCF/bbl. Typical catalysts are cobalt molybdate on alumina and nickel-tungsten on alumina for HDS and HDN, respectively, HDN requiring more severe conditions.

REFERENCE

1. Stein, T. R., Upgrading of Coal Liquefaction Products to Gas Turbine Fuels, presented at Contractors Meeting, the Electric Power Research Institute, Palo Alto, California, May 26, 1976.

Chapter 13

Upgrading of Liquid Products from the Coalcon Process

13.1 INTRODUCTION

In 1975, Coalcon Co., an affiliate of Union Carbide and Chemico, received a \$237.2 million contract from the Energy Research and Development Administration (then the Office of Coal Research) to build a coal-to-clean-liquid-fuels demonstration plant. The project concerned the design, construction and operation of a 2600-ton day^{-1} demonstration plant using a hydrocarbonization process for producing 3990 bbl day^{-1} of liquid product and 22 million ft^3 day^{-1} of pipeline quality gas.

The original plan called for two phases devoted to the design of the demonstration plant to be funded by ERDA. Phases three and four, construction of the plant and its operation through mid-1983, were to be co-funded by ERDA and private industry. However, as scale-up troubles developed in 1976, ERDA decided to terminate its support in mid-1977.

The process involves pyrolysis of coal at a hydrogen pressure of 550 psig. The coal is heated in the hydrocarbonization reactor producing liquid, gaseous, and solid products. The gas is separated from the liquid to produce synthetic natural gas (SNG) and liquified petroleum gas (LPG). The liquid is segregated into a light and heavy fraction with characteristics similar to those of a light to heavy petroleum naphtha and a light to medium gas oil, respectively. The solid char is used for fuel and for hydrogen generation.

The Coalcon process, while based on known process steps, is a relatively new project, and little has been published concerning product quality and process operation. For this reason the Coalcon process is treated in this chapter separately, while three other better-known related processes are discussed in Chapter 14.

Sufficient product data exist for the Coalcon process to define the catalytic steps needed for upgrading its liquid products. This chapter describes the conceptual design and operation of the Coalcon process, the nature of the liquid products obtained, the catalytic problems expected in upgrading these liquids, and the possible solutions of the problems.

13.2 PROCESS DESCRIPTION

Coal containing 10% or less ash and 2.9% sulfur will be stockpiled at the mine and delivered to the hydrocarbonization plant by conveyor at the rate of 15,000 tons day^{-1}. It will then be dried and sized to 50% through 100 mesh. While the Coalcon process was originally developed for low-rank western coals of low sulfur but high moisture, oxygen, and ash content, recent emphasis has been on the use of high volatility, high sulfur, and medium ash coals. Pittsburgh No. 8, Illinois No. 6, and Western Kentucky No. 11 are considered particularly attractive coals.

After preheating to 617°F, the coal is fed to the hydrocarbonization reactor as a dense fluid phase from lock hoppers which operate at 900 psig.

Hydrogen from the hydrogen plant is mixed with recycled hydrogen, preheated, and fed into the hydrocarbonization reactor to fluidize the coal. The hydrocarbonization reaction takes place at 1040°F and 550 psig. Hydrocarbon vapors and gas leave the top of the reactor, and char is discharged at the bottom.

The char exits through lock hoppers after being cooled to 600°F and stripped of hydrocarbons. It is ultimately cooled to about 200°F, pulverized and fed either to the steam boilers or to the hydrogen plant where H_2 is produced via partial oxidative gasification of the char.

The vapors which leave the top of the hydrocarbonization reactor are cooled and condensed into heavy fuel oil and light hydrocarbons. The residual gases are treated by conventional technology to remove ammonia, hydrogen sulfide, and carbon dioxide. The dried process gases are then separated cryogenically. Hydrogen is recycled to the reactor, and a CO–H_2–CH_4 stream is fed to a methanation plant to make high-Btu pipeline gas (SNG).

13.3 NATURE OF THE PRODUCTS

The Coalcon process produces four primary fuel products:

Liquefied petroleum gas (LPG)	7.3%
Synthetic natural gas (SNG)	44.7%
Light oil product	14.3%
Heavy oil product	33.7%

A model has been developed to treat pilot plant results for eastern coals [1] and to identify the nature of the product composition which is given in Table 13-1 for the hydrocarbonization reactor. It will be noted that the product slate involves very simple molecules. This may reflect an inadequacy in the extrapolation procedure or in the model itself. Tables 13-2 and 13-3 show the composition of the heavy and light oil fractions.

TABLE 13-1

POSTULATED COMPOSITION OF THE HYDROCARBONIZATION REACTOR PRODUCT IN THE COALCON PROCESS[a]

Component	Percent	Component	Percent
Hydrogen	7.66	Pyridine	3.18
Nitrogen	1.92	*p*-Xylene	0.43
Carbon monoxide	8.08	Propylbenzene	0.31
Methane	13.32	Mesitylene	1.27
Ethane	6.59	Decane	3.35
Carbon dioxide	5.38	Phenol	1.39
Hydrogen sulfide	1.90	*o*-Cresol	1.97
Organic sulfur compounds		Naphthalene	2.30
(carbonyl sulfide,		3,5-Xylenol	0.77
carbon disulfide,		1-Methylnaphtalene	0.77
thiophene)	0.35	2-Naphtol	1.61
Ammonia	0.53	3-Methylphenanthrene	1.65
Propane	3.07	1,2-Benzofluorene	4.11
n-Butane	1.05	Chrysene	10.80
n-Pentane	0.37	Char (MAF basis)	0.10
Cyclopentane	1.24	Ash	0.03
Benzene	0.65	Hydrochloric acid	0.04
Water	11.92		
Toluene	1.89		
		Total	100.0

[a] From Coals and Crelon [1].

TABLE 13-2
COMPOSITION OF THE HEAVY OIL PRODUCT FROM THE COALCON PROCESS[a]

Compound	Amount in liquid product	
	Weight percent	Mole percent
PNA fraction		
Paraffins		
Decane	0.5	0.06
Others	0.03	0.111
Total paraffins	0.53	0.171
Naphthenes		
Cyclopentane	0.6	0.144
Aromatics		
Benzene	0.06	0.118
Toluene	0.20	0.401
p-Xylene	0.10	0.149
Propylbenzene	0.10	0.152
Mesitylene	0.70	0.973
Naphthalene	8.6	10.89
Methylnaphthalene	2.9	3.28
Methylphenanthrene	6.2	5.22
1,2-Benzofluorene	15.4	11.52
Chrysene	40.4	28.70
Total	74.66	61.40
Heteroatom fraction		
Oxygenated compounds		
Phenol	1.8	3.16
o-Cresol	7.1	10.66
3,5-Xylenol	2.8	3.67
2-Naphthol	6.0	6.79
Total oxygenated compounds	17.7	24.28
Sulfur compounds		
Thiophene	0.03	0.06
Nitrogen compounds		
Pyridine	6.4	18.0
Miscellaneous		
Ash	0.9	—
H_2O	0.03	0.20

[a] From Coals and Crelon [1].

TABLE 13-3

COMPOSITION OF THE LIGHT OIL PRODUCT FROM THE COALCON PROCESS[a]

Compound	Amount in liquid product	
	Weight percent	Mole percent
PNA fraction		
Paraffins		
Methane	0.009	0.05
Ethane	0.067	0.22
Propane	0.080	1.93
n-Butane	0.89	1.53
n-Pentane	1.15	1.60
Decane	32.36	22.66
Total paraffins	34.556	27.99
Naphthenes		
Cyclopentane	8.22	11.68
Aromatics		
Benzene	6.11	7.78
Toluene	18.27	19.76
p-Xylene	4.03	3.78
Propylbenzene	2.78	2.30
Mesitylene	10.80	8.95
Total aromatics	41.99	42.58
Heteroatom fraction		
Oxygenated compounds		
Phenol	3.35	3.55
o-Cresol	0.56	0.52
3,5-Xylenol	0.33	0.27
Total oxygenated compounds	4.24	4.34
Sulfur compounds		
Thiophene	1.69	2.00
Nitrogen compounds		
Pyridine	7.74	9.74

[a] From Coals and Crelon [1].

13.3.1 Heavy Oil Product

The heavy oil fraction from the hydrocarbonization reactor has an approximate initial boiling poinpoint of 400°F and a final boiling point of 840°F which corresponds to the boiling range of a light and medium gas oil petroleum fraction with some contribution from the heavy gas oil fraction. The constituents may be conveniently divided into two broad groups, the paraffin–naphthenic–aromatic

(PNA) fraction and the heteroatom fraction (cf. Table 13-2). The hydrocarbon fraction, representing 75.8% of the heavy oil product, contains 98.5% aromatics consisting primarily of condensed two- to four-ring compounds and a very small amount of benzene and its homologs.

The heavy oil product also contains 17.7% of phenols and naphthols, 6.4% of pyridine, and 0.03% of thiophene, containing the heteroatoms O, N, and S, respectively. Although this distribution may be correct, it is likely that other oxygen, nitrogen, and sulfur compounds are also present. It is to be noted that the presence of 0.09 wt% ash in the heavy oil fraction could cause processing problems by deposition on the catalyst.

13.3.2 Light Oil Product

The light oil fraction boils in the 97–377°F range which corresponds to the boiling range of a light to heavy petroleum naphtha. Its composition is summarized in Table 13-3.

Compared to the heavy oil product, the light oil is less aromatic. The PNA hydrocarbon fraction makes up 84.8 wt% of the total light oil product and of this 40.7 wt% is paraffinic, 9.7 wt% naphthenic, and 49.6% wt% aromatic. The major component of the paraffins is decane. The only naphthene identified is cyclopentane and this is present at a relatively low level (8 wt%) compared to most petroleum naphtha fractions.

The nonhydrocarbon portion accounts for 13.7 wt% of the total light oil with 31 wt% of this fraction being phenols, 12 wt% thiophene, and 57 wt% pyridine.

13.4 UPGRADING OF THE COALCON LIQUIDS

Since few data have been reported for Coalcon liquid upgrading, anticipated problems must be extrapolated from the reported chemical and physical properties of Coalcon liquids, and by analogy to processing of similar coal liquids or petroleum fractions.

13.4.1 Heavy Oil Product: General Problems of Upgrading

The heavy liquid fraction (bp 400–840°F) is similar in physical and chemical characteristics to No. 6 fuel oil. In terms of the common refined fuels discussed in Section 12-1, one can consider upgrading this feed to gasoline, stationary turbine fuel, and refined (low sulfur and nitrogen) heating oils. Because of the high aromatic character of this feedstock, its conversion into jet and diesel fuels requiring substantial amounts of paraffinic components is not practical. However, a more severe treatment could convert it into a light aromatic blending stock for use in jet and diesel fuels. Upgrading to gasoline, turbine fuel, and refined heating oils presents two basic problems: selective cracking of the high molecular weight polynuclear aromatic constituents to lighter aromatics, and reduction of the heteroatom content to an acceptable level.

The highly condensed aromatic nature of the preponderant portion of this liquid product is a particular problem since the condensed ring compounds must be selectively cracked to one- and two-ring compounds to reduce molecular weight and thereby give improved burning characteristics. For a gasoline fraction, selective cracking would lead to high-octane constituents (BTX). In any cracking process, high levels of polycyclic aromatics promote carbon deposition and consequent catalyst deactivation. This is less of a problem for hycrocracking than for catalytic cracking because the catalysts used in the former process contain a hydrogenation function which improves activity maintenance by hydrogenating coke precursors. Selective cracking is also important to minimize light gas (C_4^-) production and maximize the liquid (C_5^+) yield.

The presence of nitrogeneous compounds at a high concentration in the heavy oil product is another problem. The nitrogen content must be reduced not only to assure an environmentally acceptable NO_x emission upon combustion but also to prevent the basic pyridine from poisoning the acidic function of the catalysts used in the upgrading of these liquids to gasoline or lighter aromatics.

The high oxygen content of the heavy liquids is also a potential problem. The presence of phenols can affect the stability, in particular during storage. However, these oxygenated compounds are likely to be removed in any hydrodenitrogenation operation.

The ash content reported to be around 0.1 wt% is definitely a problem for a fixed-bed process in which all of the ash can be deposited on the catalyst. It could also affect the use as turbine fuel (see Section 13.4.2). It may be necessary to reduce the ash level by about one order of magnitude for a continuous catalytic upgrading step.

13.4.2 Heavy Oil Product: Specific Uses

The heavy oil fraction is best suited for upgrading to refined heating and fuel oils, motor gasoline, and turbine fuels.

Refined heating and fuel oils would require some cracking and hydrocracking to reduce the molecular weight of the heavy liquid fraction and to give the appropriate viscosity, pour point, flash point, and gravity, but the first upgrading step would have to be denitrogenation.

Conversion of the heavy oil fraction to motor gasoline would require hydrocracking and denitrogenation to a product which contains mostly one- and two-ring aromatics. The presence of substantial amounts of polycyclic aromatics (primarily 1,2-benzofluorene and chrysene) is likely to promote carbon deposition during catalytic cracking. It will be difficult to selectively crack these large aromatic molecules to single-ring aromatics with minimum gas (C_4^-) production.

The requirements for upgrading the heavy oil fraction to turbine fuel can be deduced from the specifications. The constraint of 1.0 wt% maximum sulfur is easily met by the heavy oil product since it contains only 0.01% S (0.03% thiophene). The hydrogen content of the oil, around 6%, is low and will have to be increased to reach 11.3%, the specification for stationary turbine fuel. It is expected that some hydrocracking would be necessary to meet the hydrogen specification and to give the appropriate specific gravity, viscosity, and Conradson carbon residue. Denitrogenation would be necessary to meet existing and anticipated NO_x standards. Finally, the stringent specifications for sodium, potassium, lead, vanadium, and calcium are pertinent in view of the 0.1 wt% ash content of the Coalcon heavy liquid product. Such an ash content could cause substantial degradation of turbine blades if it included the mentioned elements. It is probable that any hydrofining operation, especially if carried in a fixed bed, would also remove some of the ash content by deposition on the catalyst, resulting very likely in deactivation.

A final option for upgrading the heavy liquid product would be to recycle hydrocracking to extinction to produce a heavy naphtha catalytic reforming feedstock, and blending stocks for jet and diesel fuels and for gasoline.

13.4.3 Light Oil Product

The light liquid fraction has a boiling range similar to that of motor gasoline. However, as shown in Table 13-3, its paraffinic component is primarily decane, its naphthenic content is low, and its aromaticity is high. The only upgrading which need be considered for this fraction is desulfurization (1.69 wt% thiophene) and denitrogenation (7.74 wt% pyridine), which should be easier for this oil than for the heavy oil product. After hydrotreating to remove the sulfur and nitrogen to adequate levels, this fraction could be blended to give high-octane gasoline. If the high level of decane presents an engine operability problem, two options are apparent. The hydrotreated light liquid fraction could be mixed with a light naphtha stock and processed through a reformer. This would convert a substantial fraction of the decane to more desirable isoparaffins and aromatics [2]. Alternatively, the light oil product could be used as an aromatic blending stock for jet fuel. The decane level (34.6 wt%) is attractive for this use because boiling point (345°) and energy density are favorable.

13.5 APPLICATIONS OF RECENT RESEARCH DEVELOPMENTS

From the previous discussion, it is clear that, regardless of the scheme used for upgrading heavy Coalcon coal liquids, two primary catalytic problems can be anticipated: selective cracking of four-ring aromatics to single-ring aromatics with minimum light gas make and minimum carbon deposition on the catalyst; and denitrogenation to acceptable levels.

13.5.1 Short-Range Developments

Upgrading the heavy liquid fraction to any of the previously discussed products requires hydrotreating, the degree of which depends upon the desired product. This may involve selective cracking of 1,2-benzofluorene and chrysene, to single-ring aromatics and hydrodenitrogenation of pyridine (see Fig. 13-1). For conversion to heating oils, only denitrogenation and minimal cracking are necessary, perhaps only scission to one- and two-ring compounds. In the following sections these problems will be discussed under the headings of activity and selectivity, and of activity maintenance.

13.5.1.1 Activity and Selectivity

Both selective hydrocracking and HDN require the use of bifunctional catalysts (see Section 7.5) having the proper hydrogenation and cracking functions. If the hydrogenation activity is too high, excessive and nonselective use of costly hydrogen occurs. If it is too low, activity maintenance is a problem. Similarly, the cracking function requires careful control, depending on the process (see Section 4.3.2). One of the primary tasks of a development program designed to improve cracking and in particular HDN catalysts is catalyst charac-

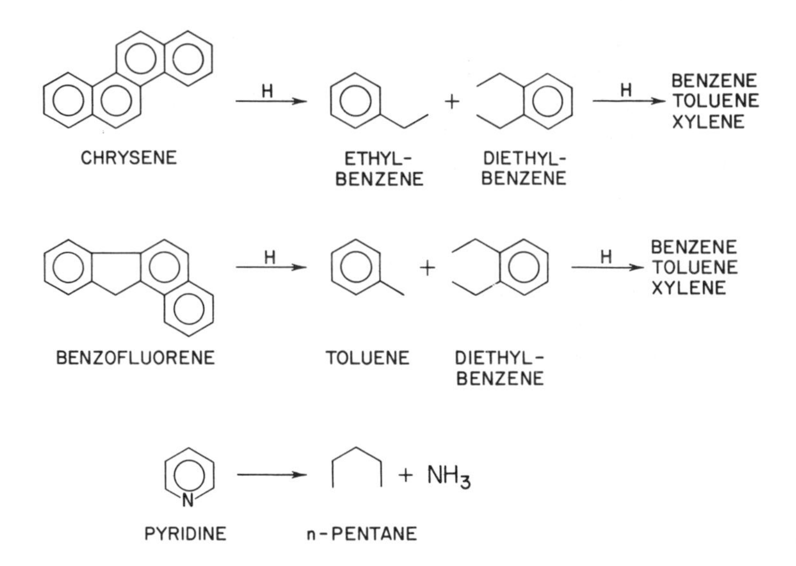

FIG. 13.1 Hydrocracking of chrysene and benzofluorene, and hydrodenitrogenation of pyridine.

terization (see Section 7.3), especially in the case of bifunctional catalysts. It is a complex problem and has not been studied in depth for such materials as supported oxides, sulfides which would be of interest for cracking, or HDN. While measurements involving the amount, type, strength, and distribution of acid sites can be carried out on a routine basis (see Section 4.2), the titration of cationic sites associated with hydrogenation activity (e.g., Cr^{3+}, Fe^{2+}, Co^{2+}, etc.) has only recently been studied by means of NO chemisorption, which should be applicable to other systems as well. Another promising adsorbate reported for measuring the number of active sites on $CoO-MoO_3-Al_2O_3$ catalysts is H_2S [3,4], but more work will be needed to develop a routine method.

In terms of catalyst improvement, the three areas which could have the most significant impact on selective cracking and HDN using existing catalysts are optimization of physical characteristics, variation of catalyst composition, and effective catalyst testing methods.

Optimizing physical characteristics involves careful control of surface area, pore volume, and pore-size distribution for conventional HDN catalysts such as Ni–W and Ni–Mo on Al_2O_3, and mixed-oxide hydrocracking catalysts such as Ni–W, Ni–Mo, or Pd supported on zeolites, $SiO_2-Al_2O_3$, TiO_2-SiO_2, and $MgO-SiO_2$ [5] (see also Chapter 3).

Control of physical properties has been recognized as an important problem for HDS catalysts and has received considerable attention in the last few years. Because of the more severe conditions required for HDN (see Section 7.5), it is expected that the optimum HDN and HDS catalysts will differ in a number of properties, including physical characteristics. Controlled pores are important to minimize diffusion limitations and to reduce catalyst deactivation. With large molecules like chrysene and 1,2-benzofluorene, such diffusion problems can readily occur and are minimized by the use of wide-pore catalysts. A number of gel procedures have recently been developed which give considerable flexibility in controlling the pore size and its distribution (see Section 5.2.2). Sometimes large pores lead to poor mechanical properties (low crush strength). Recent work has shown that this drawback can be remedied in some instances by proper doping with alkaline and rare-earth cations [6]. More work needs to be done to show the general applicability of this technique.

Variation in catalyst composition is an important effort in maximizing the activity of existing catalysts. The need for this optimization can be seen from the fact that MoO_3/Al_2O_3 and $CoO \cdot MoO_3/Al_2O_3$ have about the same activity for the HDN of pyridine [7]. This shows that, while needed for HDS, Co is not an essential constituent for HDN, except perhaps for activity maintenance. Composition can be varied readily in a systematic manner using many of the recently developed procedures including gel precipitation, sol–gels, cogels, aerogels, and homogeneous pH change (see Table 5-1). The cogel procedure reportedly gives catalysts with unprecedented activity and activity maintenance for HDS and

HDN of high molecular weight feedstocks [8]. Examples for this type of catalysts are $NiO \cdot WO_3$, $NiO \cdot WO_3 \cdot ZrO_2$, and $NiO \cdot WO_3 \cdot TiO_2$, all cogelled with ultrastable zeolites (see Section 5.2.2).

Testing catalysts for selective hydrocracking or HDN should be done in continuous units and should use model compounds (i.e., chrysene and 1,2-benzofluorene) as well as actual Coalcon heavy liquid products. Conventional trickle bed testing leads to a number of problems (see Section 8.3), and two-phase upflow systems can give results limited by mass transfer [9]. An effective reactor system for studying both general reaction variables as well as kinetics has been described by Carberry and later modified by other workers (see Section 8.3). This system is basically an autoclave with continuously flowing gas and liquids either over a spinning basket containing the catalyst or over a fixed catalyst bed flushed with gas and liquid by impeller motion.

13.5.1.2 Activity Maintenance

One cause for loss of catalytic activity is carbon deposition on the catalyst. The factors affecting carbon deposition, particularly in processing feeds such as the heavy Coalcon liquid products, include the proper balance of hydrogenation and cracking activities (see Sections 7.3 and 7.5), control of pore size and its distribution to optimize the residence time of large molecules, and maximizing the contact between catalyst, hydrogen, and liquid feed. The ratio of hydrogenation to cracking activity can be adjusted by changing the relative amounts of the transition metal oxide and solid acid components. The control of pore-size characteristics has been discussed in conjunction with catalyst activity and selectivity. Optimum contacting can be readily achieved in laboratory experiments by using the Carberry reactor (see Section 8.3). Commercial operation presents some difficulty, especially when using the not very efficient trickle bed reactor. Ebullient bed reactors as used in the H–Oil process (see Section 14.2) have some advantages.

The second cause for activity loss is thermal degradation. The two primary methods of minimizing or preventing such degradation are cationic doping of the support [6] and the use of catalyst–support interactions (see Section 3.1.2). Cationic doping with alkaline and rare-earth ions has been used to prepare aluminas which retain a surface area of 50 $m^2\ gm^{-1}$ at 1250°C [6]. Ultrastable zeolites have been synthesized which can withstand the same temperature without loss of surface area [10]. Similar effects have been reported for properly prepared mixed-oxide systems (see Table 5-1).

The use of catalyst–support interactions makes it possible to prepare catalysts for use in catalytic combustion at temperatures up to 1400°C without loss of activity [11,12]. Many of the common HDN or hydrocracking constituents such as Ni–Mo, Ni–W, and Co–Mo should be amenable to similar stabilization, as indicated by the results obtained with the catalysts prepared by the cogel process (see Section 5.2.2) and with the molybdenum-oxide–magnesia system leading to Mo_3 clusters [13].

13.5.2 Long-Range Developments

Longer-range advances in the HDN or selective hydrocracking of Coalcon heavy liquids can be expected in three areas: new synthesis procedures, new catalytic materials, and reaction studies.

13.5.2.1 New Synthesis Procedures

Some of the new procedures described in Chapter 5 allow the preparation of novel materials in high enough surface area for characterization and catalytic testing. For example, the aerogel procedure developed by Teichner (see Section 5.2.2) has been used to prepare mixed-oxide solid acids such as $NiO \cdot Al_2O_3$ and $NiO \cdot MoO_3$ with surface areas as high as 600 m^2 gm^{-1}. Such high surface area allows easy doping for chemical and structural modification. Other procedures which will be particularly valuable in preparing new solid acids and mixed transition metal oxides and sulfides include the cogel, gel precipitation, and homogeneous pH change techniques (see Section 5.2.2).

13.5.2.2 New Materials

Heavy Coalcon oil products contain only 0.01% sulfur, and therefore many potential catalysts are available for their denitrogenation and selective cracking. In addition to mixed transitional metal oxides, novel materials such as borides, carbides, nitrides, and mixed systems (e.g., Mo_2BC) are of interest (see Sections 9.4 and 9.5). Oxides containing metallic clusters of pertinent atoms (e.g., Mo_3 in $Mg_2Mo_3O_8$ and $Co_2Mo_3O_8$), not yet explored for HDN and cracking, are also potential candidates for testing [13].

13.4.2.3 Reaction Studies

The elucidation of the mechanisms of HDN and of the selective cracking of polynuclear aromatics could contribute to the development of processes for upgrading Coalcon liquids.

In the mechanism of the HDN reaction both C–N scission and hydrogenation of the nitrogen heterocycle [14] can be rate determining as shown by the difference in the mechanisms of the HDN of quinoline and indole (see Section 7.3.1). According to the data available, pyridine is the only nitrogen heterocycle in Coalcon liquids. This probably is a result of the oversimplified model used to extrapolate western coal data to those expected for eastern coals [1]. With varying coals and process conditions it is expected that some other heterocyclic nitrogen compounds will also form (e.g., carbazole, indole, and quinoline). It is particularly important to find ways to prepare HDN catalysts with significantly improved activity, as HDN catalysts can go through a maximum activity with increasing temperature [14]. This is so because higher temperatures favor dehydrogenation of heterocyclic nitrogen compounds, and in the dehydrogenated form the C–N scission occurs much less readily than in the saturated molecule.

There is some evidence that the presence of H_2S increases the rate of HDN [14,15]. Further study is needed to elucidate this effect and related interactions of ammonia, water, and oxygen with HDN.

The use of an active hydrogen donor such as tetralin in conjunction with a catalyst to hydrodesulfurize coal liquids has shown promising results. Few attempts to adapt this process for HDN have been unsuccessful so far [16], but further efforts in this direction appear warranted.

The use of supercritical conditions during the catalytic reaction (see Section 8.3) would be of interest because of the possibility of a significant increase in the hydrogen solubility in the fluid, a parameter that is important in both HDN and hydrocracking.

There is a need for studying the mechanism of selective hydrocracking of polycyclic aromatics such as chrysene and 1,2-benzofluorene. The steps in cracking a four-ring molecule like chrysene involve hydrogenation, isomerization and, cracking (see Section 7.5). Center-ring cracking is rarely observed, but some success has been reported for the cracking of phenanthrene over $Cr_2O_3 \cdot Al_2O_3$ [17] which should be studied further.

13.5.3 Supporting Research

Work in two areas would be helpful in furthering the development of the upgrading of Coalcon liquids: one concerns the analysis of the products obtained from various coals at different operating conditions; the other, the characterization of HDN and hydrocracking catalysts by the new powerful instrumental techniques [5].

13.6 CONCLUSIONS

Two areas in which catalysis could advance the upgrading coal liquids are the development of improved processes for hydrodenitrogenation and for selective hydrocracking of polycyclic aromatic molecules to single-ring aromatics. The light liquid fraction would require HDS and HDN and possibly some hydrocracking of the paraffinic constituents depending upon use. Motor gasoline and an aromatic blending stock for diesel and jet fuels are possible products. More significant HDN and hydrocracking problems are expected in upgrading the heavy liquid product. This fraction is best suited for upgrading to refined (low sulfur and nitrogen) heating and fuel oils and to stationary turbine fuels. With substantial selective hydrocracking, it can also be used as motor gasoline and as an aromatic blending stock for jet and diesel fuels.

In short-range developments, emphasis should be put on improving existing HDN and hydrocracking catalysts. This involves making use of recently developed catalyst synthesis procedures (e.g., cogel, aerogel, and homogeneous pH change) to improve the physical properties of known catalysts. Pore volume, pore-size distribution, crush strength, and thermal stability, are among the properties which can be improved by these procedures. These modifications are also expected to be useful for optimizing the activity maintenance of existing catalysts. In carrying out this work it is imperative that techniques for measuring the specific surface properties of oxides be further developed to permit the characterization of these catalysts and the determination of their intrinsic activity.

In longer-range developments, many new materials should be tested. These include new mixed oxides, oxides containing clusters of metal atoms, borides, carbides, and nitrides. Some of these may offer the possibility of modifying the interaction between heterocyclic nitrogen atoms and the surface and of improving HDN activity. The availability of many mixed-oxide materials with varying surface acidity properties are going to be important in the development of new solid acid catalysts which give selective center-ring cracking of polycyclic aromatic molecules. The mechanism of both HDN and cracking needs to be investigated.

Advances in catalytic reactor design should be utilized for testing catalysts, particularly in three-phase systems occurring in HDN or hydrocracking. The continuous-flow Carberry reactor and its modifications offer one of the most effective means for testing catalysts for these reactions. This type of reactor should also be tested in conjunction with operations at supercritical conditions, and with the use of hydrogen donors in the presence of catalysts. These approaches may provide improved contacting for multiphase systems and better access for hydrogen to the catalyst surface.

REFERENCES

1. Coals, E. T., and Crelon, C. R., Commercial Plant Process Evaluation Report (Abridged), prepared for the Energy Research and Development Administration, Fossil Energy, Coal Conversion and Utilization Division, Demonstration Plants, by Coalcon Co., New York, New York, Contract E(49–18)-1736.
2. Emmett, P. H. (ed.), "Catalysis," Vol. VI, p. 509. VanNostrand–Reinhold, Princeton, New Jersey, 1958.
3. Dillimore, D., Galevey, A., and Rickett, G., *J. Chim. Phys. Phys.-Chim. Biol.* **72**, 1059 (1975).
4. Kolosov, A. K., Shvets, V. A., and Kazanskii, V. P., *Kinet. Katal.* **16**, 197–201 (1975).
5. Boudart, M., Cusumano, J. A., and Levy, R. B., New Catalytic Materials for the Liquefaction of Coal, Electric Power Research Institute, Rep. No. RP-415-1, October 30, 1975.

6. Gauguin, R., Graulier, M., and Papee, D., *in* "Catalysts for the Control of Automotive Pollutants" (J. E. McEvoy, ed.), p. 147. American Chemical Society, Washington, D.C., 1975.
7. Sonnemans, J., and Mars, P., *J. Catal.* **31**, 209 (1973).
8. Kittrell, J. R., U.S. Patent No. 3,639,271 (1972); U.S. Patent No. 3,558,471 (1971); U.S. Patent No. 3,536,606 (1970); U.S. Patent No. 3,536,605 (1970); U.S. Patent No. 3,535,227 (1970).
9. Satchell, D. P., Development of a Process for Producing an Ashless, Low-Sulfur Fuel from Coal, Vol. IV, ERDA Contract No. E(49-18)-496, R & D Rep. No. 53, Interim Rep. No. 15, May, 1974.
10. Kerr, G. T., *J. Phys. Chem.* **73**, 1780 (1969).
11. Pfefferle, W. C., U.S. Patent No. 3,928,961 (1975).
12. Arons, R., U.S. Patent No. 3,554,929 (1971).
13. Tauster, S. J., *J. Catal.* **26**, 487 (1972).
14. Mayer, J. F., PhD Thesis, Massachusetts Institute of Technology, Cambridge, Massachusetts, 1974.
15. Goudriaan, F., Gierman, H., and Flugter, J. C., *J. Inst. Pet. London* **59**, 40 (1973).
16. Doyle, G., *Am. Chem. Soc. Div. Pet. Chem. Prepr.* **21**, 165 (1976).
17. Wu, W., and Haynes, H. W., Jr., *Am. Chem. Soc., Div. Pet. Chem. Prepr.* **20**, 446 (1975).

Chapter 14

Refining of Coal Liquids from the COED, H-Coal, and Synthoil Processes to High-Octane Gasoline, Jet Fuel, and Diesel Fuel

The first three sections of this chapter describe the nature of the COED, H-Coal, and Snythoil processes and liquids and the refining problems connected with the production of gasoline and of jet and diesel fuels from these liquids. The following two sections discuss those advances in catalysis and related disciplines which could improve the processing technology for coal liquids.

14.1 COED LIQUIDS

The operation of the COED process has been described in detail [1-11]. Coal is pulverized, dried, and then heated to successively higher temperatures in a series of fluidized-bed pyrolysis reactors in which a fraction of the volatile matter in the coal is released. Thermal staging prevents agglomeration of the coal, since in each reactor the temperature is kept just below the agglomeration point. Usually four stages are required, operating at 600, 850, 1000, and 1500°F; but the staging conditions can be adjusted to fit the type of coal processed. The heat for each of the first three reactors is provided by burning char in the fourth reactor.

TABLE 14-1
TYPICAL PROPERTIES OF COED PYROLYSIS LIQUIDS[a]

	Coal source	
	Utah A seam	Illinois No. 6 seam
Properties of Derived Oil[b]		
Carbon (wt%)	83.8	79.6
Hydrogen (wt%)	9.5	7.1
Nitrogen (wt%)	0.9	1.1
Sulfur (wt%)	0.4	2.8
Oxygen (wt%)	5.0	8.5
Ash (wt%)	0.3	0.9
H/C atom ratio	1.36	1.07
API gravity (60°F)	−3.5	−4
Moisture (wt%)	0.5	0.8
Pour point (°F)	100	100
Viscosity (SUS 210°F)	390	1333
Solids (wt%)	3.8	4.0
Gross heating value (Btu lb^{-1})	16,100	15,050

[a]From Scotti *et al* [1], courtesy of FMC Corporation and the American Institute of Chemical Engineers.
[b]All percentages on dry basis.

The volatile oil released by the pyrolysis process is condensed and filtered to remove char fines and ash. A typical inspection for the pyrolysis oil is given in Table 14-1 for Utah and Illinois coals. The pyrolysis oil is then hydrotreated at rather severe conditions (see Table 14-2) in a fixed-bed reactor to remove sulfur, nitrogen, and oxygen to produce a 25–30° API synthetic crude oil product, the syncrude.

TABLE 14-2
TYPICAL HYDROTREATING CONDITIONS FOR ILLINOIS NO. 6 SEAM COED PYROLYSIS LIQUIDS[a]

Catalyst	NiMo on alumina extrudates
Pressure	1750–2500 psig
Temperature	700–800°F
Space velocity	0.3–0.6 lb oil (hr lb catalyst)$^{-1}$
Gas recycle rate	40–80M SCF/bbl
Gas recycle concentration	90–95% H_2
Hydrogen consumption	3500 SCF/bbl

[a]From Scotti *et al* [1], courtesy of FMC Corporation and the American Institute of Chemical Engineers.

14.1.1 The Nature of COED Liquids

Typical data for COED syncrudes derived from western and eastern coals are given in Table 14-3. Some differences in the composition of the syncrude are apparent for the two coals. High-volatility bituminous Illinois coal gives a more aromatic syncrude and much less paraffins than western Utah coal. Both crudes are highly naphthenic in nature. The western coal also has a larger amount of high molecular weight fraction than the eastern coal as shown by the greater fraction boiling above 650°F.

A chemical analysis for the COED syncrude derived from Illinois No. 6 seam coal is given in Table 14-4. It is apparent that the fixed-bed hydrofining of the pyrolysis liquid increases the H/C atom ratio to 1.58, and decreases the sulfur content to 0.068 wt%, the nitrogen content to 324 ppm, and the oxygen content to 0.64 wt%.

Analysis of the syncrude fraction for two samples of an Illinois No. 6 seam coal is given in greater detail for the initial boiling point (IBP) to 390°F fraction, the 390–650°F fraction, and the 650°F$^+$ fraction in Tables 14-5, 14-6, and 14-7,

TABLE 14-3

Typical Data for COED Syncrudes Derived from Western and Eastern Coals[a]

	Coal source	
Data	Illinois Peabody mine, No. 6 seam, HvbC rank	Utah King mine, A seam, HvbB rank
Hydrocarbon type analysis (liquid, vol%)		
Paraffins	10.4	23.7
Olefins	Nil	Nil
Naphthenes	41.4	42.2
Aromatics	48.2	34.1
API gravity (60°F)	28.6	28.5
ASTM distillation (°F)		
Initial boiling point (IBP)	108	260
50% distilled	465	562
End point (EP)[b]	746	868
Fractionation yields (wt%)		
IBP–180°F	2.5	0
180–390°F	30.2	5.0
390–525°F	26.7	35.0
525–650°F	24.4	30.0
650–EP	16.2	30.0

[a] From Project COED [10].

[b] 95%, except for Illinois No. 6, which is 98%.

TABLE 14-4

ANALYSIS OF COED SYNCRUDE DERIVED FROM ILLINOIS NO. 6 SEAM COAL[a]

General specifications	
Gravity (°API at 60°F)	28.6
Flash point (°F)	70
Pour point (°F)	−55
Carbon residue (Ramsbottom) (wt%)	0.20
Hydrogen (wt%)	11.30
Carbon (wt%)	85.65
Sulfur (wt%)	0.068
Nitrogen (ppm by wt)	324
Oxygen (wt%)	0.64
H/C (atom ratio)	1.58
Hydrocarbon composition	
Paraffins	10.4
Cycloparaffins	41.4
Monoaromatics	34.1
Polynuclear aromatics	14.1

[a] From Project COED [10].

TABLE 14-5

ANALYSIS OF THE IBP–390°F FRACTION OF COED SYNCRUDE FOR TWO DIFFERENT SAMPLES DERIVED FROM ILLINOIS NO. 6 SEAM COAL[a]

	Composition (vol%)	
	Sample 1[b]	Sample 2[c]
Paraffins	7.1	6.3
Monocycloparaffins	58.9	66.4
Dicycloparaffins	12.3	12.8
Tricycloparaffins	0.2	0.2
Alkylbenzenes	19.8	13.1
Indans/tetrahydronaphthalene	1.7	1.2
Total saturates	78.5	85.7
Total aromatics	21.5	13.5
Total polyaromatics	1.7	1.2

[a] From Project COED [10].
[b] API gravity (60°F), 43.8.
[c] API gravity (60°F), 43.3.

respectively. The IBP–390°F fraction (Table 14-5) is primarily a mixture of light and heavy naphthas and is highly naphthenic, containing 70–79 vol% mono- and dicycloparaffins.

The 390–650°F fraction (Table 14-6) contains 27–30% naphthenes, over 50% of polynuclear aromatics, particularly concentrated in indan, indene, tetrahydronaphthalene, acenaphthene, and acenaphthylene structures (see Fig. 14-1).

The 650°F+ fraction (Table 14-7) contains about the same total wt% of paraffins as the two previous fractions. The total saturates level is 30–35%, and the total aromatics content has increased to 64–70%, the polynuclear aromatics representing 16–37%. It is apparent that significant differences can occur in the composition of the liquid fraction obtained from the same coal seam, but the reason for such difference is not clear.

COED liquids derived from Utah [12] and Kentucky coals [13] have recently been analyzed in even greater detail with respect to specific chemical compounds. The results of these studies generally agree with the data presented above in that the hydrocarbon types in the syncrudes were similar to, but more

TABLE 14-6

ANALYSIS OF THE 390–650°F FRACTION OF COED SYNCRUDE FOR TWO DIFFERENT SAMPLES DERIVED FROM ILLINOIS NO. 6 SEAM COAL[a]

	Composition (vol%)	
	Sample 1[b]	Sample 2[c]
Paraffins	10.9	8.0
Monocycloparaffins	11.1	14.3
Dicycloparaffins	10.1	8.9
Tricycloparaffins	5.9	7.3
Alkylbenzenes	9.4	9.6
Indans/tetrahydronaphthalene	17.1	16.1
Indenes	17.3	22.0
Naphthalene	0.3	0.0
Naphthalenes	4.6	5.2
Acenaphthenes	7.3	4.5
Acenaphthylenes	5.4	3.1
Tricyclic aromatics	0.6	1.0
Total saturates	38.0	38.5
Total aromatics	62.0	61.5
Total polyaromatics	52.6	59.6

[a] From Project COED [10].
[b] API gravity (60%F), 22.5.
[c] API gravity (60%F), 21.7.

TABLE 14-7

ANALYSIS OF THE 650°F+ FRACTION OF COED SYNCRUDE FOR TWO DIFFERENT SAMPLES DERIVED FROM ILLINOIS NO. 6 SEAM COAL[a]

	Composition (wt%)	
	Sample 1[b]	Sample 2[c]
Total paraffins	10.9	5.8
Monocycloparaffins	11.3	16.5
Polycycloparaffins	7.7	13.4
Total monoaromatics	33.0	48.6
Diaromatics	13.7	9.2
Triaromatics	8.9	2.4
Tetra-aromatics	9.6	2.7
Penta-aromatics	4.9	1.4
Total saturates	29.9	35.7
Total aromatics	70.1	64.3
Total polyaromatics	37.1	15.7

[a] From Project COED [10].
[b] API gravity (60%F), 11.2.
[c] API gravity (60%F), 13.0.

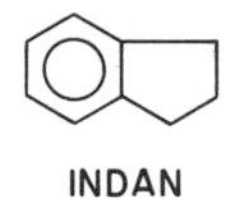
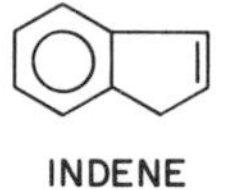
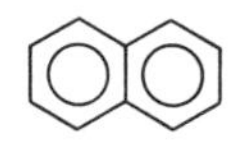

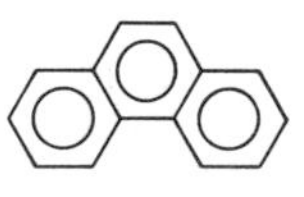

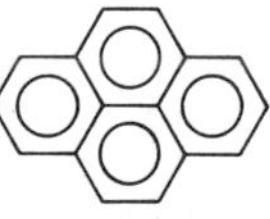

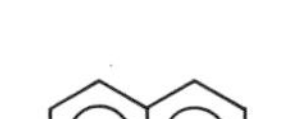
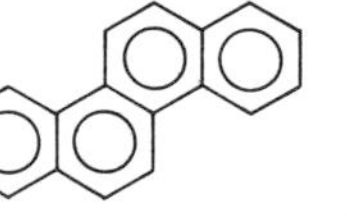
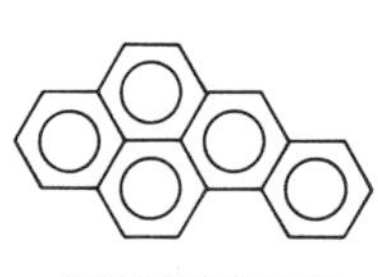

FIG. 14.1 Polycyclic aromatic hydrocarbons.

aromatic and naphthenic in character than, those in petroleum crudes. The syncrudes were also found to contain substantial amounts of oxygenates and had a uniform distribution of nitrogen compounds which ranged from pyridine to complex carbazole structures. The sulfur compounds included heterocyclic aromatic structures.

14.1.2 Refining Problems

The refining problems of COED syncrude derived from Illinois No. 6 seam coal were studied by Atlantic Richfield Co. (ARCO) in a subcontract from FMC Corp. [10] and covered hydrogen pretreatment and reforming of naphtha fractions; evaluation of middle-distillate oils as jet, diesel, and home heating fuels; fluidized-bed catalytic cracking of the atmospheric bottoms fraction; and hydrocracking of the atmospheric bottoms fraction.

14.1.2.1 High-Octane Gasoline

The COED naphtha (180–390°F) containing 70% naphthenes, 10% paraffins, and 20% aromatics is easy to process since the naphthenes readily isomerize and/or dehydrogenate over conventional reforming catalysts (e.g., Pt/Al_2O_3) to high-octane aromatics. However, because of the presence of sulfur (100 ppm) and nitrogen (100 ppm), the naphtha must be hydrotreated to reduce each of these catalyst poisons to less than 2 ppm.

The highly naphthenic nature of these naphthas ensures good activity maintenance during reforming, even for conventional Pt/Al_2O_3 catalysts. The ARCO study [10] indicates that catalyst aging for COED naphthas occurs half as fast as for petroleum naphthas if measured by the decline of the research octane number of the reformate. Also, the yield of reformate gasoline blending stock is substantially higher than from conventional petroleum naphthas. For example, operating at conditions which yield a 100 RONC product, a typical petroleum naphtha gave 75.5 vol% yield of C_5^+ fraction, compared to 91 vol% for the COED naphtha.

The major problems are encountered in the refining of the 650°F$^+$ fraction to high-octane gasoline. Prior to fluid cracking or hydrocracking this fraction, it is necessary to lower its polyaromatics and nitrogen content to avoid catalyst deactivation by excessive coke deposition and poisoning by nitrogen bases. This is done by rather severe hydrotreatment at 0.6 weight hourly space velocity (WHSV), 700°F, 2000 psig, and a hydrogen rate of 6000 SCF/bbl. The gasoline yield from fluid catalytic cracking of the hydrotreated COED atmospheric residuum is 51%, which compares favorably with a 54% yield from a mid-continent petroleum gas oil.

The difficulty in hydrocracking this 650°F$^+$ fraction to gasoline and middle distillates arises from two sources. The reduction of the nitrogen content from 800 ppm to less than 10–20 ppm required to prevent catalyst poisoning cannot be accomplished even with very severe HDN (0.33 WHSV, 750°F, 2000 psig, and 6000 SCF/bbl H_2 flow rates) because of the presence of very refractory N-heterocycles. The second source of processing problems is the presence of four- and five-ring aromatics, which adsorb very strongly on the catalyst surface and cause carbon deposition and decrease catalyst life. The only way the 650°F$^+$ fraction could be processed by ARCO was to fractionate it into a 650–800°F cut (84%) and an 800°F$^+$ cut (16%). The latter is reportedly a reasonable blending stock for No. 6 fuel oil because of its low sulfur level and reasonable Conradson carbon and polynuclear aromatics level [10]. Hydrocracking the 650–800°F fraction gives a normal product yield distribution, but requires twice as much hydrogen as needed by a corresponding petroleum feedstock and involves excessive catalyst deactivation. Therefore it appears that conventional hydrocracking technology would not be practical for COED heavy liquids.

TABLE 14-8

COMPARISON OF COED JET FUEL FRACTION WITH JP-5 JET FUEL SPECIFICATIONS[a]

Properties	JP-5	COED jet fuel (390–525°F)	
		As distilled	2-Stage HDS[b]
ASTM 10% distillation point (°F)	400 max.	436	412
ASTM distillation end point (°F)	550 max.	530	526
Gravity (°API, 60°F)	36–48	25.7	33.2
Sulfur (wt%)	0.4 max.	0.007	0.001
Freezing point (°F)	−51 max.	<−70[c]	<−80[c]
Net heating value (Btu lb^{-1})	18,300 min.	18,070[c]	18,390[c]
Aniline gravity product[d]	4500 min.	1490	4501
Aromatics (liquid vol%)	25 max.	60.7	0–5
Olefins (liquid vol%)	5 max.	0	0
Smoke point[e]	19 min.	10	22
Luminometer number[f]	50 min.	20.4[c]	43.6[c]
Flash point	140 min.	186	156
Copper corrosion test[g]	1	1	—

[a] From Project COED [10].
[b] After hydroprocessing of 390–525°F syncrude fraction.
[c] Calculated.
[d] A measure of the aromaticity of an oil.
[e] A measure of the smoke and soot-producing characteristics of jet fuels.
[f] A measure of the flame temperature of jet fuels.
[g] A measure of the corrosiveness of oil to copper; a "1" designation indicates slight tarnish.

14.1.2.2 JET FUEL

The COED middle distillate can be processed to both jet and diesel blending stocks. However a highly aromatic product such as derived from Illinois No. 6 seam coal needs hydrotreating before it can be used for this purpose.

As shown by Table 14-8, the "as-distilled" 390–525°F fraction does not meet the primary jet fuel specifications for a number of parameters including gravity, aromatics content, smoke point, luminosity, and net heating value [10]. In an attempt to overcome some of these limitations the fraction was hydrotreated at 600–700°F, 1000–2000 psig, 1.0 WHSV, and 5000 SCF/bbl H_2 flow rate. However, even at these conditions the product could not meet all specification, including smoke point, probably because of the very high cycloparaffin content. The only way in which the smoke point specification could be met was by using a second stage noble metal (Pt) hydrogenation at 700°F, 2000 psig, 5000 SCF/bbl H_2 flow rate, and 2 WHSV to reduce the aromatic level to zero. It is expected that all specifications will only be met by selective cracking of the cycloparaffins to brânched and linear paraffins at low temperatures and high hydrogen pressure to prevent the formation of aromatics. It is questionable whether this would be a

TABLE 14-9
COMPARISON OF A COED DIESEL FUEL FRACTION WITH ASTM SPECIFICATIONS[a]

		As distilled			
Properties	Specifications	390–650°F	525–650°F	390–525°F	390–525°F Fraction 2-stage HDS[b]
Pour point (°F)	0 max.	−70	−20	<−80	<−80
Flash point (°F)	130 min.	215	300	186	156
Sulfur (wt%)	Legal	<.001	<.001	.007	<.001
ASTM distillation (°F)					
10% distilled	460 max.	459	562	436	412
50% distilled	—		588	455	437
90% distilled	620 max.	586	626	494	483
100% distilled	660 max.	613		530	
Gravity (°API, 60°F)	33–37	22.5	19	25.7	33.2
Cetane number[d]	30 min.	18	24	23.0	>30[c]
Aniline point (°C)[e]	—			14.3	58.2
Aromatics (liquid vol%)	—			58.3	5.2

[a] From Project COED [10].
[b] After hydroprocessing of 390–525°F fraction of syncrude.
[c] Estimated.
[d] A measure of the ignition quality of diesel fuel.
[e] A measure of the aromaticity of oil.

cost-effective process because with less difficulty it is possible to convert this fraction into an aromatic feedstock for gasoline blending.

14.1.2.3 Diesel Fuel

The 525–650°F COED fraction is highly aromatic and therefore does not meet diesel specification because of high density (19° API versus 33–37° API) and low cetane number [10].

A 390–650°F fraction can be distilled from the middle-distillate cut which meets diesel fuel specifications for flash point, distillation, and pour point, but the high aromaticity still gives an unacceptably low gravity and cetane number. On hydrotreatment the gravity specification is met, but the distillation end point is low. These data are presented for comparison in Table 14-9.

The problems in converting COED syncrude to a diesel fuel are, therefore, similar to those for producing a jet fuel. It might be possible to hydrotreat the middle-distillate fraction to a paraffinic–cycloparaffinic fuel which would just meet product specifications, but the most desirable normal and branched paraffins can only be made by selective hydrogenation and hydrocracking requiring substantially greater amounts of hydrogen.

14.2 H-COAL LIQUIDS

The H-Coal process is based on the H-Oil process which was developed several years ago by Hydrocarbon Research, Inc. (HRI) for hydrotreating heavy petroleum feedstocks [14–16]. The key aspect of the process is the ebullient reactor into which is fed a slurry of hydrogen, coal, coal liquids, and catalyst. The catalyst is promoted cobalt molybdate with appropriate attrition and pore-size characteristics. The coal is hydrogenated and converted to liquid and gaseous products. Careful sizing of the coal and catalyst particles permits unconverted coal, ash, liquids, and gases to leave the reactor without loss of catalyst from the reactor. Catalyst is usually added or withdrawn continuously to maintain constant activity.

An advantage of the ebullient bed is the highly effective contacting of coal, hydrogen, liquids, and catalyst in this multiphase system.

14.2.1 The Nature of H-Coal Liquids

Depending upon process conditions the H-Coal process produces either a syncrude or a low-sulfur fuel oil. Typical compositions of both product types are given in Table 14-10 for the liquefaction of Illinois No. 6 seam coal. The conversion and product yield are about the same whether a syncrude or fuel oil is produced. However, the latter involves a lower hydrogen consumption and there-

TABLE 14-10
PRODUCT DISTRIBUTION AND YIELD FOR H-COAL LIQUIDS DERIVED FROM ILLINOIS NO. 6 SEAM COAL[a]

	Product	
	Synthetic crude	Low-sulfur fuel oil
Normalized product distribution		
C_1–C_3 hydrocarbons	10.7	5.4
C_4–400°F distillate	17.2	12.1
400–650°F distillate	28.2	19.3
650–975°F distillate	18.6	17.3
975°F$^+$ residual oil	10.0	29.5
Unreacted ash-free coal	5.2	6.8
H_2O, NH_3, H_2S, CO, CO_2	15.0	12.8
Total (100.0 + H_2 reacted)	104.9	103.2
Conversion (%)	94.8	93.2
Hydrogen consumption (SCF/ton)	18,600	12,200
Product yield		
Wt% of M.A.F. coal	74.0	78.2
Barrels per ton of M.A.F. coal		
C_4–400°F fraction	1.32	0.90
400–650°F fraction	1.64	1.14
650–975°F fraction	0.98	0.89
975°F$^+$ residue	0.44	1.39
Total	4.38	4.32

[a]From Johnson, courtesy of the Society of Mining Engineers.

TABLE 14-11
PRODUCT INSPECTION OF C_4^+ LIQUIDS FROM H-COAL PROCESSING OF ILLINOIS NO. 6 SEAM COAL[a]

	Product	
	Synthetic crude	Low-sulfur fuel oil
°API gravity	15.0	4.4
Hydrogen (wt%)	9.48	8.43
Sulfur (wt%)	0.19	0.43
Nitrogen (wt%)	0.68	1.05

[a]From Johnson, courtesy of the Society of Mining Engineers.

TABLE 14-12

COMPOSITION OF C_4–400°F FRACTION FOR H-COAL LIQUIDS FROM ILLINOIS NO. 6 SEAM COAL[a]

Component	Wt%	Component	Wt%
Paraffins		Alkylbenzenes	
nC_4	0.10	C_6	0.89
iC_5	0.20	C_7	3.77
nC_5	0.69	C_8	4.76
C_6	2.48	C_9	4.16
C_7	2.87	C_{10}	2.58
C_8	2.08	C_{11}	1.29
C_9	1.59	C_{12}	0.10
C_{10}	1.19		
C_{11}	0.69		17.55
C_{12}	0.10		
	11.99		
Saturated naphthenes		Other aromatics	
Monocycloparaffins	42.64	Indans	6.44
Dicycloparaffins	8.50	Naphthalenes	0.59
Tricycloparaffins	0.19		
			7.03
	51.33		
Unsaturated naphthenes		Phenols	
Monocycloolefins	5.32	Mol. wt. 108	0.13
Dicycloolefins	4.98	Mol. wt. 122	0.56
Tricycloolefins	0.90	Mol. wt. 136	0.19
		Mol. wt. 150	0.02
	11.20		
			0.90
		Total =	100.00 wt%

[a] From Project H-Coal Report [17].

fore gives a higher molecular weight product. The sulfur and nitrogen levels in the syncrude product (Table 14-11) are lower than those in the fuel oil.

The compositions of various fractions distilled from the H-Coal syncrude are given in Tables 14-12 to 14-14. The C_4–400°F or naphtha fraction is highly naphthenic and contains 7% polynuclear aromatics and 0.9% phenols (Table 14-12). The middle-distillate (400–650°F) fraction contains almost 60% aromatics which are predominantly polynuclear in character (Table 14-13), 2% phenols, and 3.1% other nonhydrocarbons. The fraction boiling at 650–919°F consists mostly of aromatics (Table 14-14), 72.8% being naphthalenes, phenanthrenes,

TABLE 14-13
COMPOSITION OF 400–650°F FRACTION FOR H-COAL LIQUIDS FROM ILLINOIS NO. 6 SEAM COAL[a]

Component	Wt%	Component	Wt%
Saturated hydrocarbons		Aromatic hydrocarbons	
n-Paraffins	4.8	Alkylbenzenes	12.6
i-Paraffins	1.7	Indans and tetralin	30.8
Monocycloparaffins	14.0	Indenes	5.7
Dicycloparaffins	7.9	Naphthalene	0.2
Tricycloparaffins	2.6	Naphthalenes	3.5
	——	Acenaphthenes (C_nH_{2n-14})	4.0
	31.0	Acenaphthylenes (C_nH_{2n-16})	2.2
		Tricyclics (C_nH_{2n-18})	0.4
			——
			59.6
Unsaturated naphthenes		Nonhydrocarbons (phenols)	
Monocycloolefins	4.3	Mol. wt. 108	0.04
	——	Mol. wt. 122	0.52
	4.3	Mol. wt. 136	0.98
		Mol. wt. 150	0.38
		Mol. wt. 164	0.07
		Mol. wt. 178	0.01
		Other nonhydrocarbons	3.10
			——
			5.10
		Total =	100.00 wt%

[a] From Project H-Coal Report [17].

chrysenes, benzanthracenes, benzphenanthrenes, and pyrenes. The nonhydrocarbon portion amounts to about 14 wt%, containing primarily sulfur and nitrogen compounds.

14.2.2 Refining Problems

Few data have been published concerning the refining H-Coal liquids on a large scale, but the key catalytic problems can be anticipated from the product composition data.

14.2.2.1 HIGH-OCTANE GASOLINE

The composition of the naphtha fraction given in Table 14-12 indicates that the conversion of this product to high-octane gasoline will involve dehydrogenation and hydroisomerization to aromatics. Both reactions occur readily over conven-

TABLE 14-14
COMPOSITION OF 650–919°F FRACTION FOR H-COAL LIQUIDS FROM ILLINOIS NO. 6 SEAM COAL[a]

Component	Wt%	Component	Wt%
Saturated hydrocarbons		Unsaturated naphthenes	
Paraffins	1.4	Olefins	0.0
Monocycloparaffins	3.1	Monocycloolefins	0.5
Bicycloparaffins	0.6	Bicycloolefins	0.3
Tricycloparaffins	0.7	Tricycloolefins	0.2
Tetracycloparaffins	0.4	Tetracycloolefins	0.2
Pentacycloparaffins	0.2	Pentacycloolefins	0.1
Hexacycloparaffins	0.1	Hexacycloolefins	0.1
Other	0.3	Other	0.2
	6.8		1.6
Aromatic hydrocarbons		Nonhydrocarbons	
Alkylbenzenes	3.0	S and N compounds	13.8
Indans and/or "Tetralins"	0.5		
Other aromatics[b]	72.8		
	76.3		

[a] From Project H-Coal Report [17].

[b] An approximate breakdown of aromatic hydrocarbons is as follows in mmoles/100 gm: naphthalenes, 93.4; phenanthrenes, 91.1; chrysenes, 21.9; 1,2-benzanthracenes, 3,4-benzphenanthrenes, 14.6; pyrenes, 15.4; 5-ring aromatics, 5.1.

tional reforming catalysts. Activity maintenance would not normally be expected to be a problem for such a highly naphthenic naphtha. However the bicyclic aromatics, 7% of indans and naphthalenes, could, because of their strong adsorption by acidic sites, cause carbon deposition with time unless the reforming is carried out under hydrogen pressures higher than used normally.

The 400–650°F fraction should be convertible to a naphthenic–aromatic naphtha by a moderate hydrotreating operation in which oxygen, sulfur, and nitrogen compounds are removed to sufficiently low levels to allow further processing either by reforming or by mixing with catalytic cracker feedstock.

Processing the 650–919°F fraction presents more serious problems because of the presence of 73% of refractory polynuclear aromatics and of 13.8% of high molecular weight nonhydrocarbons which are difficult to desulfurize and denitrogenate. One possible option would be to distill off a 650–800°F fraction for processing with the 400–650°F fraction. The remaining heavy residuum could then be used as a blending stock for heavy fuel oils if the sulfur, nitrogen, and polynuclear aromatic constraints are met.

14.2.2.2 Jet Fuel

The middle and heavy H-Coal liquids are highly aromatic and therefore do not yield a good jet fuel. It is likely that severe hydrotreating would improve the product quality for certain specifications sufficiently to allow the product to be used as a jet fuel blending stock.

14.2.2.3 Diesel Fuel

The 400–650°F fraction from the H-Coal process is highly aromatic and therefore is not expected to meet diesel specifications, particularly for density, cetane number, and oxidation stability. Severe hydrotreating and more effective hydrocracking will be needed to upgrade H-Coal liquids to a diesel fuel.

14.3 SYNTHOIL LIQUIDS

In the Synthoil process coal is slurried in a recycled aromatic oil and flowed, together with hydrogen, at highly turbulent conditions through a fixed bed of cobalt molybdate catalyst [18–25]. Depending upon the conditions of operation, either a heavy or a light low-sulfur fuel oil is produced. Coal conversion levels of 90–98% are common [18]. As an example, a Kentucky coal with 5.5% sulfur and 16% ash can be converted to an oil which flows at room temperature and has 0.2 wt% sulfur, 0.8 wt% nitrogen, and 0.2 wt% ash. Process conditions are usually either 2000 or 4000 psig at 450°C, depending upon whether the desired product is a light or heavy oil. Usual contact time is 2 min. Oil yields are about 3 bbl/ton of coal with a hydrogen consumption of 4000–5000 SCF/bbl of oil, depending upon product quality.

A unique feature of this process is that the ash in the coal causes controlled attrition of the catalyst surface thereby minimizing deposition of mineral matter and carbon. The ash is ultimately separated from the coal liquids by centrifugation.

14.3.1 The Nature of Synthoil Liquids

A recent study by Woodward *et al*. [25] reports on the characteristic constituents of Synthoil liquids produced from a West Virginia Pittsburgh seam coal at 4000 psig and 450°C by passing hydrogen gas and a 35 wt% coal slurry in recycle oil at 25 lb h^{-1} through a 14.5-ft-long Synthoil reactor containing ⅛-in. pellets of cobalt molybdate on silica-stabilized alumina. The product was a brownish-black liquid with a specific gravity of 1.081, containing 0.79 wt% of nitrogen and 0.42 wt% of sulfur. Comparison of the Synthoil syncrude with that produced by the COED process from Utah coal [26] and western Kentucky coal [27] showed the Synthoil product to be much heavier, and containing a consider-

TABLE 14-15

COMPARISON OF DISTILLATES (wt%) OF SIMILAR BOILING RANGE FRACTIONS FOR THREE SYNCRUDES PRODUCED BY THE SYNTHOIL AND COED PROCESSES[a]

	Synthoil, West Virginia			COED			
				Western Kentucky[b]		Utah[b]	
Coal type boiling range (°F)	405°F$^-$	405–685°F	685–988°F	400–716°F	716°F$^+$	400–716°F	716°F$^+$
Saturates	30.3[c]	16.0	9.7	25.0	23.8	27.8	25.8
Monoaromatics	27.6	27.3	4.7	42.0	25.1	25.1	14.4
Diaromatics	3.2	21.6	22.6	13.0	24.3	17.5	18.4
Polynuclear aromatics		7.9	41.1	5.4	20.0	7.1	25.1
Heteroatomics	35.3	22.2	15.6	4.4	4.5	15.2	7.4
Distillate wt% of syncrude		42.6	27.3	54.2	24.2	45.4	40.3

[a] From Woodward, courtesy of Bartlesville Energy Center.
[b] Neither Western Kentucky nor Utah COED syncrudes contained appreciable material boiling above 988°C.
[c] Includes 3.2% olefins.

ably greater amount of carbon residue in both the residuum and total crude oil [25]. The synthoil product had higher nitrogen content and substantially more sulfur. The nitrogen and sulfur compounds of the Synthoil liquids are concentrated in the heavy ends.

Additional data comparing Synthoil coal liquids and COED coal liquids from three different coals are given in Table 14-15. Although the starting coals vary in physical and chemical properties, some useful general points can be made. The West Virginia and western Kentucky coals are somewhat similar in overall properties. The concentration of saturates is lower in the middle and heavy Synthoil fractions compared to the corresponding values for the COED liquids. The concentration of monoaromatics in the 685–988°F fraction of the Synthoil liquids is also lower than in the similar COED fractions. The 685–988°F Synthoil fraction contained 64% aromatics with more than one ring while the similar COED fractions contained 43–44%. For comparison, it may be noted that the corresponding petroleum distillate fractions contain 17–21% polynuclear aromatics [25].

These differences are mentioned to indicate the diversity of composition which can be found among the syncrudes produced from various coals using various coal liquefaction process. It would be of interest, in this respect, to liquefy a given seam of coal using the various conversion processes to ascertain the exact differences in composition due primarily to processing variations.

The physical properties of three primary fractions boiling below 405°F, 405–685°F, and 685–988°F are summarized in Table 14-16.

The 405°F$^-$ fraction contained 35.3 wt% acids which were almost entirely phenols, 27.1% paraffins and cycloparaffins, and 30.8% aromatics.

TABLE 14-16

Physical Property Data for Three Distillates from West Virginia Coal Liquid[a]

Property	<405°F Distillate	405°–685°F Distillate	685°–988°F Distillate
Specific gravity at 60°/60°F	0.936	0.990	1.109
Gravity (°API)	19.7	11.4	
Pour point (ASTM D-97, °F)	<5	<5	—
Color (BuMines description)	Brownish black	Brownish black	Greenish black
Kinematic viscosity (100°F, ASTM D-445, Cs)	2.27	9.56	—
Saybolt viscosity (SUS, 100°F)	34	57	—
Sulfur (bomb) (ASTM D-129, wt%)	0.20	0.30	0.44
Nitrogen (Kjeldahl) (wt%)	0.423	0.724	1.187
Carbon residue (Conradson) (ASTM D-524, wt%)	1.29	2.33	7.42

[a]From Woodward, courtesy of Bartlesville Energy Center.

The 405–685°F and 685–988°F fractions were separated into specific compounds representative of three broad classes of compounds: monoaromatics, diaromatics, and polynuclear aromatic-polar materials. The monoaromatics consisted of a single benzene ring with several (up to six) naphthene rings attached. The average carbon number was C_{13} (with a range of C_{10}–C_{28}) for the lower boiling fraction and C_{22} (with a range of C_{15}–C_{30}) for the high boiling fraction.

Diaromatics in the 405–685°F fraction consisted of naphthalenes, diphenylalkanes, and their naphthenologs with an average C_{13} carbon number (with a range of C_{10}–C_{26}). In the 685–988°F fraction diaromatics were primarily naphthalenes with up to nine naphthene rings attached to the naphthalene nucleus. The average carbon number was C_{20} (with a range of C_{15}–C_{34}).

The polynuclear aromatic-polar group of the 405–685°F distillate was chromatographically separated into neutral hydrocarbons (10.3%), weak acids (7.2%), strong acids (9.1%), and bases (3.4%). The hydrocarbon fraction consisted of acenaphthylenes, phenanthrenes, anthracenes, diacenaphthylenes, pyrenes, benzopyrenes, perylenes, chrysenes, and their naphthenologs (see Fig. 14-1).

The weak acid compounds were determined to be indoles, quinolines, and carbazoles, as well as some oxygenates and sulfur compounds. The strong acids were found to be phenols and naphthenophenols, diaphthenophenols, and indenols. The base fraction consisted primarily of pyridines, acridines, and quinolines, the latter being the major group.

The 685–988°F fraction was separated into hydrocarbons and weak acids (48.4%), strong acids (6.8%), and bases (4.7%). The hydrocarbons were found to be naphthenologs of acenaphthylenes, phenanthrenes, anthracenes, and chrysenes.

The strong and weak acids were not readily spearated for the high boiling fraction, but oxygenates and such nitrogen compounds as alkylcarbazoles were detected. The high boiling fraction contained 2.6% nitrogen compounds and 1.4% sulfur compounds. The bases extracted from this fraction were primarily quinolines and acridines.

14.3.2 Refining Problems

Four potential problem areas are expected to affect the required processing steps of Synthoil liquids: high Conradson carbon in all fractions; high concentrations of large polynuclear aromatic molecules; the presence of heterocyclic sulfur, nitrogen, and oxygen compounds; and the ash content.

14.3.2.1 High-Octane Gasoline

Before the Synthoil 405°F$^-$ fraction can be used as a naphtha feedstock it requires severe hydrotreating to convert the high concentration of phenols

(35.3%) to single-ring aromatics, to reduce the nitrogen (0.423%) and sulfur (0.20%) levels to less than 2 ppm, and to minimize the high level of Conradson carbon (1.29%).

The 405–685°F fraction contains almost 30% di- and polynuclear aromatics which are basic in nature and adsorb strongly on metals and oxides and lead to excessive catalyst deactivation by blocking active sites, as well as by acting as coke precursors. This middle-distillate fraction could be hydrocracked or catalytically cracked, but would also require hydrotreatment to lower the sulfur (0.30%) and nitrogen (0.724%) levels and to reduce the Conradson carbon (2.33%). An alternate possibility would be to distill off a lighter fraction and blend the heavy residuum with fuel oil.

The heavy fraction, 685–988°F, is even more difficult to process than the middle distillate because of the higher sulfur, nitrogen, and Conradson carbon levels. The di- and polynuclear aromatics content is 63%, most of the constituents containing large condensed structures highly substituted with naphthenes. Unless this fraction is subjected to a very severe hydrotreatment, it is likely that it would have to be vacuum distilled and the residuum coked.

14.3.2.2 Jet Fuel

The middle distillate can potentially be converted to jet fuel, or at least to jet fuel blending stock, but would require pretreatment. The primary problem is likely to be the high level of aromatics. The Synthoil fraction contains 27% monoaromatics, 21% diaromatics, and 8% polynuclear aromatics, substantially more than the level of 25 vol% allowed for jet fuel. High-pressure hydrotreating could saturate many of these aromatic structures to naphthenic materials and thus improve smoke point and energy density. However, in the presence of substantial amounts of naphthenes it may be necessary to reduce the aromatics level to practically zero, as their smoking properties are enhanced in the presence of naphthenes. Ultimately, the most effective processing for converting this fraction to jet fuel would be hydrogenation to naphthenic structures followed by selective cracking of these saturates to branched paraffins and paraffin-substituted single-ring aromatics.

14.3.2.3 Diesel Fuel

Because of its highly aromatic nature, the Synthoil middle-distillate fraction would not meet the cetane number and specific gravity requirements for diesel fuels (see Table 14-9).

High-pressure hydrotreating and selective cracking to a paraffinic–naphthenic fuel is expected to be necessary to reduce aromaticity and produce a fuel which meets the specifications.

14.4 IMPACTING AREAS AND RECENT DEVELOPMENTS

Although the syncrudes produced from a given coal may vary for the COED, H-Coal, and Synthoil processes, it is apparent that the compositions are at least qualitatively similar with respect to various molecular constituents. Therefore, the technical difficulties in converting these syncrudes to high-octane gasoline, jet fuel, and diesel fuel are expected to be similar and to vary primarily in degree. The COED and H-Coal processes generally give a more naphthenic naphtha which is easier to reform than that which is obtained from Synthoil syncrudes. However, both H-Coal and Synthoil naphthas contain polynuclear aromatics and heterocycles which leads to catalyst deactivation—an especially severe problem for Synthoil liquids.

The high aromaticity of the various syncrudes makes them difficult to process, particularly to paraffinic jet and diesel fuels. The molecular compositions of the starting syncrudes and of the three distillate fractions indicate that there are three major technical requirements: selective hydrocracking of polynuclear aromatics, more effective hydrotreating processes (particularly for HDN), and the development of catalysts with improved activity maintenance.

14.4.1 Selective Hydrocracking of Polynuclear Aromatics

Distillation of COED, H-Coal and Synthoil liquids into light naphtha, middle distillate, and residuum oil gives fractions with substantial amounts of molecules containing three to nine condensed aromatic rings, sometimes highly substituted with naphthenes or paraffins. It is essential to selectively hydrocrack these multi-ring structures with a minimum consumption of hydrogen, as they lead to catalyst deactivation in many conventional petroleum upgrading processes. For the production of jet and diesel fuels the rings must be either converted to cycloparaffins or, more preferably, the latter hydrocracked to branched paraffins.

Unfortunately, only a limited amount of work has been done concerning detailed kinetic studies of the catalytic conversion of these large ring structures, primarily because of the experimental difficulties connected with rapid catalyst deactivation and with reliable product analyses. Over the last decade the development of chromatographic and spectrographic techniques [28] has permitted some progress in the analytical area. Pertinent studies have included the hydrocracking of naphthalene, anthracene, phenanthrene, and pyrene (see Section 7.5). It is most desirable to crack the polynuclear aromatics selectively to substituted benzenes with minimum light gas yield. In the case of phenanthrene, it is preferable to saturate the center ring and subsequently to crack only this part of the molecule to form ethyl- and methylbenzenes. This has been observed with limited selectivity on catalysts consisting of alumina with controlled acidity and of valence-stabilized chromia [29]. The use of this type of catalyst in an improved form may be an approach to enhancing the selectivity for center-ring cracking reactions.

Another approach is provided by the finding of Qader *et al.* [30] that CoS mechanically mixed with $SiO_2 \cdot Al_2O_3$ is more active for hydrocracking aromatic ring structures than either component by itself. This observation indicates a bifunctional mechanism for hydrocracking and the occurrence of migrations of reactants or intermediates from one catalyst component to the other. These workers also found a marked effect of the Si/Al ratio, high ratios giving more active catalysts. CoS appeared to be better than NiS, WS_2, and MoS_2 for minimizing coke formation, possibly because of its higher hydrogenation activity. All of this suggests that low-temperature hydrocracking catalysts will require an optimum ratio of the hydrogenation and cracking functions.

There is a need for very active catalysts for cracking the large rings, as lower temperatures favor cracking over condensation to coke precursors. If higher temperatures are required, it may be necessary to quench free-radical intermediates which initiate condensation. One way to do this is to use hydrogen donors which selectively transfer hydrogen atoms to these intermediates. Some work in this area has recently been reported [31] in conjunction with donor desulfurization of heterocyclic aromatics.

Advances in catalyst preparative techniques allow the preparation of a variety of oxides, sulfides, oxysulfides, and other complex materials, with controlled physical and chemical properties (see Section 5.2). Pore-size distribution, surface area, and acidity, are primary parameters for the hydrocracking of large aromatic molecules. Preparative procedures such as gel precipitation, homogeneous pH change, aerogels, pyrogels, and sol–gels are useful for preparing mixed oxides like $TiO_2 \cdot SiO_2$, $Al_2O_3 \cdot MgO$, and $ZrO_2 \cdot TiO_2$ with varying pore-size distribution and acidity. It is expected that many of these techniques will be important for synthesizing new catalysts.

With large molecules as condensed polycyclic aromatics and heterocyclics, it is crucial to have wide pore catalysts to facilitate their access to the active sites. As these pores are made larger by various preparative procedures, the mechanical strength of the catalyst begins to deteriorate rather quickly. Recent work has indicated that this flaw might be remedied to some degree [32] in the case of alumina by doping with alkaline and rare-earth cations. It is not clear whether the resulting improvement in mechanical properties is due to the formation of a more rigid lattice network or to enhanced adhesion between particles which make up the alumina structure in the catalyst pellet.

It is apparent that the hydrogenation to cracking activity ratio is important for selective hydrocracking. Interesting combinations of such activities may be provided by the recently discovered novel inorganic materials. These include complex oxides which may contain both catalytic functions, e.g., zeolites, perovskites like $SrTiO_3$ (perhaps with modified acidity by addition of another oxide like SiO_2 or Al_2O_3); bronzes like $NaWO_3$; oxides containing metal–metal bonded clusters like $Mg_2Mo_3O_8$ or $Co_2Mo_3O_8$; a large number of complex sulfides such as $BaZrS_3$ (perovskite), $FeCr_2S_4$ (thiospinel), and $Al_{0.5}Mo_2S_4$; and a host of carbides (e.g., W_2C, Pt_3SnC, $W_{16}Ni_3C_6$), nitrides (e.g., Co_3N_2, V_2N, Ti_2AlN, Ni_3AlN), borides (MoB, $Co_{21}Hf_2B_6$, Ni_3B), phosphides (Co_2P, FeP_2), and silicides (Mo_3Si, TiSi) (see Chapter 9).

Depending upon the upgrading step and the level of sulfur in a given fraction, these materials will vary in chemical stability. Particularly influenced by a sulfiding environment are metals and multimetallic systems, although thermodynamic calculations suggest that the ultrastable Engel–Brewer intermetallic compounds such as $ZrPt_3$ may have enhanced sulfur resistance at low to moderate sulfur levels (see Section 10.2).

Numerous new catalyst preparative procedures have been developed which allow the preparation of some of these materials with sufficiently high surface area for catalytic testing (see Section 5.2). Complex oxides, sulfides, and oxysulfides are of particular interest because the ratio of hydrogenation to cracking activity can be fine-tuned for these materials by controlling composition (see Section 3.1). It appears that sulfur sensitivity can be affected in some cases. Zeolites in combination with amorphous mixed oxides have significant potential. Work by Dalla Betta *et al.* [33] has shown that by careful preparative procedures one can put a highly dispersed transition metal in various parts of the zeolite network. By this procedure it is possible to enhance the sulfur tolerance of Pt atoms in certain zeolites because there is an electron withdrawal from the Pt atoms which apparently weakens any interaction with sulfur. This is supported by the fact that interactions with bases like NH_3 are enhanced. Indeed, preliminary work with zeolites in combination with hydrogenation functions such as WS_2 or MoS has shown that they have potential for hydrocracking polynuclear aromatics to single-ring structures (see Section 7.5).

In carrying out catalytic studies with model polynuclear aromatic structures it is essential to work with well-characterized catalysts. At minimum, this means that the total surface area, average pore diameter, and possibly selective adsorption measurement of the active catalytic surface are required (see Chapter 4 and Section 13.5.1).

The hydrocracking of the middle distillate, and particularly of the heavy ends from any of the syncrudes discussed in this chapter, is, by the very nature of the system, carried out in liquid phase. In addition to deactivation phenomena, contacting of the catalyst with hydrogen and the large polynuclear aromatic structures is a problem and inevitably leads to mass-transfer limitations. More effective contacting can be obtained by the use of homogeneous catalysts (see Section 9.6), molten salts (see Section 9.7), and supercritical conditions (see Section 8.3). Homogeneous catalysis, while conceptually promising, is fraught with the difficulties of catalyst loss and recovery. Molten salts have similar problems, although in recent work improved catalyst recovery to the 99.9% level for direct coal liquefaction has been reported [34]. The probability of success using melts for hydrocracking heavy feedstocks is substantially greater than for coal conversion since loss due to eutectic formation with mineral matter is decreased. The concept of supercritical catalysis is relatively new, but preliminary data suggest that this technique offers a number of possible advantages for hydrocracking heavy fractions. For example, direct extractions of coal at 660°F and 1450 psi with toluene gave products which indicated very limited degradation at these relatively severe conditions (see Section 8.3). In the catalysis area the Lewis-acid-catalyzed isomerization of paraffins [35] using a CO_2–hydrocarbon solvent at 1000–5000 psig and temperatures up to 200%C showed a fivefold increase in the ratio of isomerization to cracking over that observed in the CO_2-free system.

14.4.2 Hydrotreating Processes

The present discussion will be restricted to processes concerned with the reduction of the level of sulfur and nitrogen compounds in syncrudes to allow further catalytic processing without the danger of poisoning. Removal of oxygenated compounds needs no special consideration. It is relatively easy, involving mostly phenols, and occurs usually in the desulfurization or denitrogenation step.

14.4.2.1 HYDRODESULFURIZATION

Hydrodesulfurization is kinetically less difficult than HDN, although substituted benzothiophenes and naphthobenzothiophenes present both steric and chemical problems for conventional HDS catalysts (see Section 7.4.2). The HDS studies carried out over cobalt molybdate catalysts on 3,7-dimethylbenzo-[b]thiophene suggest that the desulfurization step involves C—S bond breaking as an initial step, and that aromatic saturation is not a prerequisite for C—S bond breaking when the carbon atom is aromatic. This suggests that the appropriate selective catalyst should be able to remove sulfur without doing unnecessary hydrogenation and thus save hydrogen costs.

There is indication that C—S hydrogenolysis and hydrogenation reactions proceed on separate sites (see Section 7.4.3), the latter occurring on strongly electrophilic sites, the former on weakly electrophilic sites. If this is true, the synthesis of more active catalysts for HDS may require a bifunctional approach. Many of the novel inorganic materials mentioned in Section 14.4.1 could find application in such studies, sulfides and oxysulfides being particularly interesting. The latter have not been tested extensively for HDS. Kolboe and Ambery (see Section 7.4.1) showed that the addition of oxygen to thiophene during HDS over MoS_2 caused a doubling of the rate of HDS. The effect of oxygen on butene (a proposed intermediate in thiophene HDS) hydrogenation is even greater, increasing it one hundredfold. This effect may be due to the formation of atomic hydrogen via OH radicals generated in the reaction of hydrogen and oxygen.

Complex oxides of the type $M_2Mo_3O_8$ (where M = Mg, Zn, Co, Mn, and others) have been suggested as HDS catalysts [36] because their Mo atoms are located at the apices of equilateral triangles with a Mo–Mo distance of 2.53 Å, which is shorter than in Mo metal. These materials are now available in a form with high surface area [36] and exhibit unusual catalytic properties for hydrogenation and hydrogenolysis reactions, with activities between those observed for metals and oxides. The study of HDS of realistic model compounds such as dibenzothiophene and even larger molecules using these and other complex oxide and oxysulfides could give some leads for new more active HDS catalysts.

According to Doyle [31], the combination of an HDS catalyst (cobalt molybdate) with a hydrogen donor such as tetrahydronaphthalene synergistically enhances desulfurization over the rates found by the use of the donor or the catalyst separately, suggesting a promising approach for improving desulfurization.

14.4.2.2 HYDRODENITROGENATION

Hydrodenitrogenation has two objectives in removing nitrogen compounds: to avoid catalyst poisoning in downstream refining steps, and to minimize NO_x emission in the combustion of the final fuel product. The latter objective is responsible for a more recent intensive interest in HDN studies.

An HDN study on quinoline suggests that C–N hydrogenolysis is the rate-limiting step (see Section 7.3.1). If this can be confirmed, it may indicate that bifunctional systems play a significant role in HDN. Thus a controlled acid function may be used to interact just sufficiently with the basic nitrogen atom, while an adjacent surface site (perhaps a transition metal cation) interacts with the α carbon atom in the nitrogen heterocycle. This effect could lead to substantial weakening of the C–N bond and perhaps to increased activity. It would permit operation of HDN at lower temperatures, where coke formation is minimal.

For feeds low in sulfur, chemical stability will not be a severe problem, and in such cases a number of metals, multimetallic systems, and especially the Brewer-type intermetallic compounds, such as $ZrPt_3$ (see Section 10.2), could find applications.

In using oxide and sulfide catalysts, the conditions for presulfidation could have a significant effect on the optimum HDN activity. Goudriaan and coworkers [37] studying pyridine HDN over cobalt molybdate catalysts found that presulfided catalyst was substantially more active than the oxide and that continuous addition of H_2S further increased activity. They concluded that the hydrogenation activity of the sulfided catalyst is substantially greater than that of the oxide and that the presence of H_2S had a beneficial effect on the hydrocracking activity of the catalyst. Similar effects were reported by Mayer [38]. More work with realistic model compounds at controlled conditions, using well-characterized systems, is needed to clarify these effects. The effects of NH_3, H_2O, O_2, CO, and Co_2 should be studied as well.

It has been known for some time that heterocyclic nitrogen compounds promote catalyst deactivation. Kinetic studies of polynuclear heterocyclic nitrogen compounds are hampered because of side reactions which reportedly lead to unstable intermediates and subsequent coking of the catalyst surface (see Section 7.3.1). This is shown by the work of Madison and Roberts [39] who studied the liquid phase pyrolysis of aromatics and related heterocyclics and found that the substitution of a nitrogen atom in a polynuclear aromatic substantially increased the coking rate. For example, acridine coked ten times faster than anthracene. The gas formed contained an equal quantity of hydrogen and methane, indicating extensive decomposition of the aromatic nucleus. Similar effects were found for quinoline and naphthalene, the latter forming no coke under the test conditions while the quinoline formed 4.9 wt% coke. These authors concluded that replacement of a CH group in naphthalene or anthracene by a nitrogen atom leads to a decrease in the dissociation energy of one or more of the C—H bonds, thereby facilitating condensation. The position of the nitrogen atom also has an effect, as shown by the fact that isoquinoline coked more rapidly than quinoline. According to these experiments nitrogen compounds not only deactivate catalysts by interacting with acidic sites but also cause enhanced coking rates relative to other polynuclear aromatics. Therefore more active catalysts operating at lower temperatures will have an unexpectedly greater effect on activity maintenance if activity loss is caused largely by coking of nitrogen heterocycles.

14.4.3 Activity Maintenance

In reforming, hydrotreating, hydrocracking, and catalytic cracking, the most important modes of deactivation include coking, sintering, poisoning by sulfur and nitrogen compounds, and mineral deposition. The problems of HDS and HDN have already been discussed in detail above. Therefore, this section will concentrate on catalyst deactivation by coking, sintering, and mineral deposition.

Coking is one of the major problems in catalyst deactivation. This is particularly true in the presence of polynuclear aromatics, and especially of those which contain nitrogen atoms. Madison and Roberts [39] have observed that the thermal pyrolysis of polynuclear aromatics occurs more readily in liquid phase than in vapor phase and yields mainly methane, indicating degradation of the ring structure. It was also found that molecular structure significantly affects coking rates. Thus, compounds having an anthracene or chrysene nucleus, such as benz[a]anthracene, naphthacene, and benzo[a]pyrene, formed coke much more readily than such aromatics as biphenyl, naphthalene, phenanthrene, triphenylene, pyrene, fluoranthrene, and decacyclene.

Szwarc and Shaw [40] showed that reactivities of aromatic molecules with $CH_3\cdot$ radicals correlated well with the energy of excitation of the aromatic nucleus from its singlet ground state to the first excited triplet state. In the latter the hydrocarbon showed diradical character and was reactive toward other radicals. Madison and Roberts confirmed this [39] by finding lower singlet–triplet excitation energies for all hydrocarbons possessing the anthracene structure and thus establishing a correlation with high coking tendency. According to these workers, the pyrolysis-coking reactions do not occur by simple polymerization, following dehydrogenation, but depend on the easy activation to diradicals, facilitating the formation of a complex of two molecules, which then can form a condensation product by elimination of hydrogen.

Two approaches can be taken to avoid catalyst deactivation by polynuclear aromatic coking. First, a more active catalyst for cracking these structures can be formulated which permits operation at low enough temperatures to minimize the thermal free-radical reactions. Second, the radicals can be quenched by hydrogen donors and by catalysis under supercritical conditions.

For processing fractions which contain substantial amounts of such coke precursors as polynuclear aromatics, it is sometimes possible to use a catalyst which either minimizes carbon deposition or can tolerate high coke levels in reforming naphtha to high-octane gasoline. Exxon's KX-130 and Chevron's Pt-Re are examples of such catalysts.

Sintering can come about by continued oxidative regeneration of catalysts or by operation at high temperatures in a reactive environment. Sintering models have been developed to explain the modes of thermal deactivation of catalysts (see Section 10.4). The pertinent sintering models indicate that catalyst stability can be enhanced by controlling support morphology and structure and by enhancing the catalyst–support interaction (see Section 3.1). In this regard, the use of structurally stabilized solid acids and of ultrastable zeolites as the acid function will be important. Similarly, novel preparative procedures yielding thermally stable catalysts (see Section 5.2.2) will be useful. Thus, the cogel method gives such catalysts as $NiO \cdot WO_3 \cdot ZrO_2$ and $NiO \cdot WO_3 \cdot TiO_2$ cogelled with ultrastable zeolites which maintain a surface area of 350 $m^2\ gm^{-1}$ at 1200°C for 2 h.

The third mode of catalyst deactivation occurs by mineral deposition. The mineral content of the syncrudes from the COED, H-Coal, and Synthoil liquefaction processes can be as high as 0.2–0.3 wt% and consists primarily of silicates, aluminates, and compounds of iron and titanium, as well as trace amounts of numerous other elements. The deposition of certain mineral components on a catalyst surface during upgrading not only blocks active sites but can also modify the chemical state of the catalyst. Thus small amounts of TiO_2 on a silica or alumina support can increase surface acidity causing increased hydrocracking to light gases and consequent liquid product losses and increased use of hydrogen.

The mineral problem is a very difficult one. Little work has been done to solve it. One program sponsored by ERDA is being performed at the Sandia Laboratories in New Mexico [41] to elucidate the deactivation of the catalyst used in the Synthoil process operated at very high linear velocities to promote controlled attrition of carbon and minerals deposited on the catalyst.

Another way to solve the mineral problem would be to use homogeneous catalysis, but this method would need, because of its limitations, a major innovation to make it practical.

The use of molten salt catalysts is still another approach and is under study by the University of Utah and by the Consolidation Coal Company [34].

14.5 CONCLUSIONS

Various fractions from COED, H-Coal, and Synthoil syncrudes can be processed to high-octane gasoline, jet fuel, and diesel fuel, using conventional petroleum technology. The processing presents difficulties of a greater or lesser degree, depending upon the fraction used and the desired product. The production of gasoline involves no major problem since this product is derived from the light naphtha fractions which have usually naphthenic or aromatic nature and contain the precursors for high-octane constituents.

The problems occur in upgrading the middle-distillate or heavy-end fractions to single-ring aromatics or to low molecular weight cycloparaffins or branched paraffins needed for jet and diesel fuels. The polynuclear aromatics and high molecular weight heterocycles contained in these fractions are difficult to convert to the desired molecules and also cause rapid catalyst deactivation in naphtha reforming, hydrocracking, and catalytic cracking. Therefore, these fractions require severe hydrotreating before they can be processed by conventional petroleum technology. Such hydrotreating is costly because, being nonselective, it uses excessive amounts of hydrogen. There are, therefore, three broad areas which require new scientific and technological development: selective hydrocracking of polynuclear aromatics, hydrotreating processes, and catalysts with improved activity maintenance.

The following sections summarize the conclusions presented in this chapter concerning desirable research efforts in these areas.

14.5.1 Selective Hydrocracking of Polynuclear Aromatics

(a) Development of new bifunctional catalysts with controlled ratio of cracking to hydrogenation activities: emphasis should be on novel inorganic materials, particularly complex oxides and oxysulfides, and on the use of novel catalyst preparative procedures to control surface acidity, pore-size distribution, crush strength, and attrition.

(b) Improvement of catalyst characterization techniques: comparison of activities of all catalysts on a rate per unit surface area basis; methods for titrating transition metal cations with gases should be perfected.

(c) Study of model reactions with polynuclear aromatic molecules present in coal liquids: investigation of the catalytic properties of novel catalysts for promoting center-ring scission and saving hydrogen.

(d) Elucidation and control of thermal free radical reactions: study of pyrolysis of polynuclear aromatics present in various fractions; development of hydrogen donor catalysis.

(e) Development of surface characterization procedures for complex oxides and sulfides.

(f) Study of improved methods for minimizing mass transfer problems, including cracking by molten salts and catalysis under supercritical conditions.

14.5.2 Hydrotreating Processes

(a) Development of bifunctional HDS/HDN catalysts: investigation of complex oxides, sulfides, and oxysulfides, and the concept of dual sites which interact with C and either S or N; variation of the ratio of hydrogenation to hydrogenolysis activity by catalyst support interactions and controlled acidity; optimization of pore-size distribution and of other physical properties.

(b) Study and optimization of the effects of presulfidation and of the addition of NH_3, O_2, H_2O, CO, and CO_2 on activity and activity maintenance of HDS and HDN catalysts.

(c) Study of HDS and HDN using model compounds (dibenzothiophene, carbazole, and acridine) actually present in coal liquids.

(d) Study of the nature of thermal coking reactions for N and S heterocycles: investigation of routes to minimize these cracking reactions by hydrogen donor effects and supercritical conditions.

14.5.3 Activity Maintenance

(a) Study of the pyrolysis of polynuclear aromatics and methods to avoid coking reactions.

(b) Improvement of the sintering resistance of catalysts and supports by catalyst–support interactions and the use of new doping procedures and catalyst preparation (e.g., cogels, ultrastable zeolites).

(c) Development of novel procedures for catalyst regeneration, including the use of reagents which selectively remove minerals.

REFERENCES

1. Scotti, L. J., Jones, J. F., Ford, L., and McMunn, B. D., "Multi-Stage Fluidized-Bed Pyrolysis of Coal at the Project COED Pilot Plant," FMC Corporation R & D Center, Princeton, New Jersey, presented at American Institute of Chemical Engineers, Pittsburgh, Penn., June 2–5, 1974.
2. Project COED, "Char Oil Energy Development," FMC Corporation, Princeton, New Jersey, Final Report, OCR Contract No. 14-01-0001-235, June 1962–December 1965.
3. Project COED, "Char Oil Energy Development," Amendment 3, Final Report, OCR Contract No. 14-01-0001-235, FMC Corporation, Princeton, New Jersey, January–October 1966.
4. Project COED, "Char Oil Energy Development," Interim Report No. 1, OCR Contract No. 14-01-0001-498, FMC Corporation, Princeton, New Jersey, September 1966–February 1970.
5. Project COED, "Char Oil Energy Development, The Desulfurization of COED Char, Part III," FMC Corporation, Princeton, New Jersey, Interim Report No. 2, OCR Contract No. 14-01-0001-498, December 1968–May 1970.
6. Project COED, "Char Oil Energy Development," Final Report, OCR Contract No. 14-01-0001-498, FMC Corporation, Princeton, New Jersey, October 1966–June 1971.
7. Project COED, "Char Oil Energy Development," Interim Report No. 1, OCR Contract No. 14-32-0001-1212, FMC Corporation, Princeton, New Jersey, July 1971–June 1972.
8. Project COED, "Char Oil Energy Development," Interim Report No. 2, OCR Contract No. 14-32-0001-1212, FMC Corporation, Princeton, New Jersey, July 1972–June 1973.
9. Greene, M. I., Scotti, L. J., and Jones, J. F., "Low Sulfur Synthetic Crude Oil from Coal," presented at the Division of Fuel Chemistry Symposium, ACS Meeting, Los Angeles, April, 1974.
10. Project COED, "Char Oil Energy Development," Interim Report No. 3, OCR Contract No. 14-32-0001-1212, FMC Corporation, Princeton, New Jersey, July 1972–June 1973.
11. Project COED, "Char Oil Energy Development," Interim Report No. 4, OCR Contract No. 14-32-0001-1212, FMC Corporation, Princeton, New Jersey, July 1972–June 1973.
12. Dooley, J. E., Sturm, G. P., Woodward, P. W., Vogh, J. W., and Thompson, C. J., Analyzing Syncrude From Utah Coal, Bartlesville Energy Research Center, RI-75/7, 1975.
13. Sturm, G. P., Woodward, P. W., Vogh, J. W., Holmes, S. A., and Dooley, J. E., "Analyzing Syncrude from Western Kentucky Coal." Bartlesville Energy Research Center, RI-75/12, 1975.
14. Johnson, C. A., Stotler, H. H., and Wolk, R. H., H-Coal Process for Producing Liquid Hydrocarbons, presented at *AIME Ann. Meeting, 103rd, Dallas, Texas* February 24–28, 1974.
15. Burke, D. P., *Chem. Week*, p. 38, September 11, 1974.
16. Johnson, C. A., and Livingston, J. Y., H-Coal: How Near to Commercialization?, Hydrocarbon Research, Inc., presented at University of Pittsburgh, School of Engineering, Symposium, August 6–8, 1974.
17. Project H-Coal Report on Process Development, Hydrocarbon Research, Inc., R & D Rep. No. 26, U.S. Dept. of the Interior, Office of Coal Research, Washington, D.C., 1967.
18. Yavorsky, P. M., Akhtar, S., Lacey, J. J., Weintraub, M., and Reznik, A. A., *Chem. Eng. Progr.* **71**, 79 (1975).

19. Friedman, S., Yavorsky, P. M., Akhtar, S., and Wender, I., Coal Liquefaction, U.S. Dept. of the Interior, Bureau of Mines, Pittsburgh, Pennsylvania, 1974.
20. Yavorsky, P. M., Synthoil Process Converts Coal into Clean Fuel Oil, U.S. Dept. of the Interior, U.S. Bureau of Mines, Pittsburgh, Pennsylvania, 1973.
21. Akhtar, S., Friedman, S., and Yavorsky, P. M. *AIChE Symp. Ser.* **70** (137), 106 (1974).
22. Yavorsky, P. M., Hydrodesulfurization of Coal into Nonpolluting Fuel Oil. Pittsburgh Energy Research Center, U.S. Bureau of Mines, October 1972.
23. Akhtar, S., Mazzocco, N. J., Weintraub, M., and Yavorsky, P. M., Synthoil Process for Converting Coal to Nonpolluting Fuel Oil, U.S. Dept. of the Interior, Bureau of Mines, Pittsburgh, Pennsylvania, 1974.
24. Akhtar, S., Friedman, S., and Yavorsky, P. M., Low-Sulfur Liquid Fuels from Coal, presented at *Symp. Quality Synthetic Fuels.* American Chemical Society, Boston, Massachusetts, April 9–14, 1972.
25. Woodward, P. W., Sturm, G. P., Jr., Vogh, J. W., Holmes, S. A., and Dooley, J. E., Compositional Analyses of Synthoil From West Virginia Coal, Bartlesville Energy Research Center, RI-76/2, 1976.
26. Dooley, J. E., Hirsch, D. E., Thompson, C. J., and Ward, C. C., Analyzing Syncrude from Utah Coal, Bartlesville Energy Research Center, RI-75/7, 1975.
27. Sturm, G. P., Jr., Woodward, P. W., Vogh, J. W., Holmes, S. A., and Dooley, J. E., Analyzing Western Kentucky Syncrude, Bartlesville Energy Research Center, RI-75/12, 1975.
28. Pichler, H., and Schulz, H., *Chem.-Ing. Tech.* **42**, 1162 (1970); Pichler, H., and Hector, A., Kirk-Othmer Encyclopedia of Chemical Technology, Vol. IV.
29. Wu, W., and Haynes, H. W., Jr., *Am. Chem. Soc. Div. Pet. Chem. Prepr.* **20**, 466 (1975).
30. Qader, S. A., *J. Inst. Pet.* **59**, 178 (1973); Qader, S. A., and Hill, G. R., *Am. Chem. Soc. Div. Fuel Chem. Prepr.* **16**, 93 (1972); Qader, S. A., Sridharan, R., and Hill, G. R., Abstracts of Papers, *North Am. Meeting Catal. Soc., 2nd, Houston, Texas* p. 75 (1971); Qader, S. A., and Hill, G. R., 160th Nat. Meeting, *Am. Chem. Soc. Div. Fuel Chem. Prepr.* **14**, No. 4, Part 1, 84 (1970).
31. Doyle, G., *Am. Chem. Soc. Dov. Pet. Chem. Prepr.* **21**, 165 (1976).
32. Gauguin, R., Graulier, M., and Papee, D., *in* "Catalysts for the Control of Automotive Pollutants" (J. E. McEvoy, ed.), p. 147. American Chemical Society, Washington, D.C., 1975.
33. Dalla Betta, R. A., and Boudart, M., *Proc. Int. Congr. Catal. 5th, 1972* **2**, 1329. North-Holland Publ., Amsterdam, 1973.
34. Liquefaction and Chemical Refining of Coal, Battelle Energy Program Rep., p. 46, July 1974.
35. Bartle, K. D., Martin, T. G., and Williams, D. F., *Fuel* **54**, 226 (1975).
36. Tauster, S. J., *J. Catal.* **26**, 487 (1972).
37. Goudriaan, F., Gierman, H., and Flugter, J. C., *J. Inst. Pet. London* **59**, 40 (1973).
38. Mayer, J. F., PhD Thesis, Massachusetts Institute of Technology, Cambridge, Massachusetts, 1974.
39. Madison, J. J., and Roberts, R. M., *Ind. Eng. Chem.* **50**, 237 (1958).
40. Szwarc, M., and Shaw, A., *J. Am. Chem. Soc.* **73**, 1379 (1951).
41. Chemical Studies on Synthoil Process, Sandia Laboratories, Albuquerque, New Mexico, Second Quarterly Rep., December, 1975–Feburary, 1976.

Chapter 15

Conversion of Solvent Refined Coal to Low Sulfur, Low Nitrogen Boiler Fuel

In the solvent refined coal process (SRC), coal is dissolved in an aromatic (anthracene) oil at 800°F under a hydrogen pressure of 1000 psi. The undissolved material, containing almost all the ash constituents and a substantial amount of inorganic sulfur, is then filtered off, and the solvent is recovered by distillation. The resulting product, SRC, has a melting point around 350°F, contains less than 1% S, and has very little ash (~0.1%).

SRC will therefore easily comply with the less than 0.5% ash requirement for boiler fuel and will probably satisfy the sulfur specification also, currently 1.2 lb SO_2 per million Btu. However, its nitrogen exceeds the 0.5% limit, and its H/C atomic ratio is below the required 1.6.

In this chapter a brief summary of the constraints and problems involved in upgrading SRC will be presented first, followed by an analysis of the new developments and concepts which may impact on these problems.

15.1 CATALYST CONSTRAINTS AND PROCESS REQUIREMENTS FOR SRC UPGRADING

Recent work by Mobil [1] has shown that SRC contains a large fraction of asphaltenes, and has the highest aromatic level of a number of coal liquids tested.

These constituents raise problems of the activity, selectivity, and activity maintenance in processing SRC.

15.1.1 Activity

In the hydrotreatment of petroleum residuum, one of the main catalyst problems is the fast initial deactivation by the deposition of carbonaceous materials [2] which is aggravated by the deposition of mineral matter throughout the reaction. In view of the composition of the residuum, the initial carbon deposition is believed to be due to the large asphaltene molecules [3] which are 40–50 Å in size, and plug the similarly sized micropores of the alumina support representing a large fraction of the total catalyst surface area.

In upgrading SRC the carbon deposition problem is likely to be at least as severe as observed for petroleum processing due to the high asphaltenic content of the coal. The ash content of SRC is also expected to have an adverse effect on catalyst stability because even a content of 0.1% represents a relatively large quantity of ash. In a pilot plant processing 100 tons of SRC per day, such a level would amount to 200 lb of ash per day, which could cause serious deactivation.

The presence of asphaltenes leads to additional problems in coal and residuum hydrotreatment processes by limiting diffusion in the catalyst pores and in the main fluid stream of the three-phase reaction. The choice of reactor systems used in catalyst testing is therefore particularly important for SRC as well as for other coal conversion studies.

15.1.2 Selectivity

In the upgrading of SRC to boiler fuels, the objective of primary importance is the minimization of hydrogen consumption necessary to adjust the H/C ratio and to remove the nitrogen compounds from SRC. The acidity of the catalyst plays an important part in these processes (see Sections 7.3.1 and 14.4.2.2). Both surface acidity and the nature of the catalyst are therefore important parameters in efforts to maximize selectivity. Most currently used liquefaction catalysts are designed for optimum HDS but they are not necessarily the best for SRC upgrading.

15.1.3 Activity Maintenance

In spite of the efforts to minimize HDS, some sulfur removal is likely to occur during SRC upgrading, yielding some H_2S, the presence of which will have to be considered in selecting catalysts and assessing their stability in a sulfiding environment.

The regeneration of SRC-upgrading catalysts may create some serious thermal stability and sintering problems because of the presence of sodium in SRC ash. This is based on the belief that sodium is the cause of irreversible surface area loss during the regeneration of hydrotreating catalysts used for petroleum residua [4].

15.2 IMPORTANT RESEARCH DEVELOPMENTS

15.2.1 Short-Term Developments

The discussion of the constraints and problems expected for SRC upgrading will cover activity and selectivity, and activity maintenance.

Before proceeding with the discussion it is useful to consider the question whether SRC upgrading could be competitive with alternative processes such as the direct liquefaction of coal to boiler fuels. In making this comparison, it is important to note the unique properties of SRC: low mineral content, chemical structure, and relatively low sulfur and high nitrogen concentrations. The effect of these properties on catalyst activity, its maintenance, and regeneration will have to be investigated in any short-term R and D effort in this area. It will be necessary, for example, to determine whether the mineral content of SRC is sufficiently low to minimize catalyst attrition as well as catalyst deactivation due to deposition of mineral matter on the coal. Similarly, it will be important to consider the chemical structure and composition of SRC, since it will determine carbon deposition on the catalyst, and therefore catalyst deactivation. Finally, it should be assessed whether the low sulfur composition of SRC would permit the use of more effective HDN catalysts than currently used in other coal liquefaction processes.

Unless these properties of SRC facilitate upgrading to boiler fuels considerably compared to direct liquefaction, the additional processing may not be warranted.

15.2.1.1 Activity and Selectivity

Hydrotreating is believed to require a bifunctional catalyst (see Section 7.4.3). The successful use of such catalysts in catalytic reforming of naphtha since the 1950s [5] is in large measure due to techniques for catalyst characterization which allow determination of the specific surface area of the various catalyst functions and the elucidation of their respective roles.

Characterization procedures for oxide catalysts used for coal conversion are more complex (see Section 4.3.2) and have received increasing attention in the last few years. A promising development in this area is the use of NO and H_2S

chemisorption [6] for the determination of the specific surface area of oxides. Refinements in this method will be important in the efforts to improve currently used HDN and liquefaction catalysts.

To minimize the initial rapid carbon deposition in hydrotreating, the optimum pore-size distribution for SRC-upgrading catalysts will have to be determined. For the preparation of support materials with controlled surface area and pore-size characteristics there are novel methods available: the gel procedures using organic gelling agents (see Section 5.2.2). For improving the mechanical strength of large-pore supports the addition of alkaline and rare-earth cations seems useful [7]. This method should be studied in more detail. Another variable that affects carbon deposition is the acidity of the support. The methods developed for acidity measurement (see Section 4.4) and for the synthesis of oxide catalysts with controlled acidity properties will contribute significantly to the improvement of coal conversion catalysts, both in terms of activity maintenance by minimizing coking, and of selectivity by reducing H_2 consumption.

Optimization of SRC-upgrading catalysts will require significant improvement in the hydrodenitrogenation activity. Mechanistic studies suggest that scission of the carbon–nitrogen bond of a heterocyclic molecule is the rate-determining step for the HDN (see Sections 7.3.1 and 14.4.2.2). In the commonly used cobalt molybdate system, there are indications that Mo is responsible for the scission activity; therefore a change in the strength of the Mo–N interaction is likely to have a significant effect on the HDN activity. The chemical interaction of the catalyst with the support (see Section 3.1.2) is expected to be pronounced for supported oxide catalysts and may be an appropriate method to vary the strength of the Mo–N interaction and thus influence HDN activity (see Section 7.3.2).

Improvement efforts should concentrate on modifications of currently used hydrotreating catalysts to optimize catalyst morphology and composition for HDN and SRC liquefaction. The need for modification is mostly clearly seen in the effect of the Co–Mo ratio on HDN activity (see Section 7.3.1) which suggests that, in contrast to HDS, the presence of Co is not important for HDN.

The use of catalyst–support interactions to stabilize a catalytic system is being reported in an increasing number of areas of catalysis and is likely to be especially effective for nonmetallic supported catalytic systems (see Section 3.1.2). The choice of constituents of the support to optimize the catalyst–support interactions is made on the basis of the solid-state chemistry of the catalytic material. Molybdenum, for example, is known to interact with alkaline earth oxides to form complex oxides such as $Mg_2Mo_3O_8$ [8], and this interaction may offer an interesting method of stabilization.

Thermal stabilization of the support is important in the improvement of existing hydrotreatment catalysts. Recent patents show that cationic doping prevents the high-temperature transformation of Al_2O_3 to the low surface area α

form [7]. This method and its extensions will stabilize HDN catalysts, particularly during oxidative regeneration. The formulation of the support to optimize catalyst–support interactions and stability should take into consideration the sulfiding tendency of the catalyst in view of the presence of H_2S in the SRC conversion.

15.2.2 Long-Term Developments

The specific problems of SRC upgrading may require a more extensive long-term program devoted to the application of new synthesis methods and to the development of new catalytic materials.

Combining the information collected in the last decade on the variation in acidic properties of mixed oxides with composition [9, p. 70] with new techniques for the preparation of high surface area mixed oxides (see Section 5.2.2) is likely to lead to new catalytic materials of interest for the conversion of coal. One procedure of special interest, the aerogel technique, has been used to synthesize mixed oxides such as $NiO \cdot Al_2O_3$ and $NiO \cdot MoO_2$ with surface areas as high as 600 $m^2\ gm^{-1}$ (see Section 5.2.2) and may permit the preparation of a number of new complex oxides.

Another gel procedure [10] involves the incorporation of various oxides into zeolites leading to remarkable thermal stability and HDN activity and activity maintenance for heavy feedstocks. The key to this formulation is the high stability of the zeolite component and the uniform and homogeneous nature of the catalyst.

One of the major factors which limits the choice of new catalysts for direct coal liquefaction is the presence of H_2S (see Boudart *et al.* [9] and Chapter 9). Because of the low H_2S level in SRC upgrading, the number of potential catalytic materials increases considerably. There are some limitations since metals and most alloys will sulfide at H_2S levels expected to be in the neighborhood of 0.1% (see Table 9-3). Possible candidate materials include transition metal borides (see Section 9.5), certain carbides (see Section 9.4), including the so-called Novotny phases containing several metal atoms, boride-carbides such as Mo_2BC (see Section 9.5), cluster compounds such as $Co_2Mo_3O_8$ [11] and $Al_{0.5}Mo_2S_4$ [12], and a number of complex nitrides which are similar to the Novotny carbides and have not before been tested for catalytic applications (see Section 9.4).

In the efforts to develop improved HDN catalysts, it will be important to continue mechanism studies. Since the current belief is that the rate-determining process is the scission of the nitrogen–carbon bond (see Section 7.3.1), an increase in the nitrogen–catalyst interaction will be important in efforts to improve HDN activity. Variations in the composition of some of the complex oxides, oxysulfides, and other compounds discussed above are likely to lead to an optimization of this property.

The identification of improved catalysts will require the testing of HDN activity with real feeds rather than with model compounds. A study of the hydrodenitrogenation of shale oil suggests that there is considerable interaction between the various nitrogen compounds present [13] and that the observed HDN activity observed for one nitrogeneous constituent may not hold in a mixture with others.

15.3 SUPPORTING RESEARCH

The equilibrium of H_2S with the catalytic material is an important factor in the development of new coal conversion catalysts. The stoichiometry of a sulfide is very sensitive to sulfur pressure [9, p. 147], and its change with conditions is likely to affect the catalytic behavior. Knowledge of the thermodynamics of sulfide formation is therefore crucial in the choice of potential new catalysts. Studies in this area should include the possibility of partial sulfidation of a number of compounds, and particularly of oxide systems, since oxysulfides are likely to be formed in any conversion system using oxidative regeneration.

Another area of interest for the general understanding of coal conversion processes is the effect of particle size. It has been shown to be important in the hydrogenation activity of WS_2 catalysts [14], and may well be significant for HDN (see Section 3.2).

The effect of particle size is related to the changes in structure and chemistry of the catalyst surface with size. A number of spectroscopic techniques have been studied in the last decade for the characterization of the surface of systems of interest in catalysis (see Section 11.1). Some of these capabilities have been applied to systems of interest to SRC upgrading and coal conversion. Measurement of surface composition of Ni–S catalysts by Auger electron spectroscopy, for example, has shown a relationship between composition and hydrogenation activity [15]. This type of relationship is likely to be important in catalytic systems exposed to H_2S, such as encountered in SRC upgrading. Information on the surface composition–activity relationship together with a knowledge of the thermodynamics of the systems will be valuable in the development of optimum catalysts. For the exploration of the chemical and electronic environment of the elements in the catalyst, ultraviolet photoelectron spectroscopy and x-ray spectroscopy have been used in the study of a number of compounds including oxides and sulfides of Mo and Co [16,17]. This application of these techniques promises to provide a useful guide for understanding the surface properties of materials of catalytic interest before and after reaction. For use under reaction conditions a technique of great interest is x-ray adsorption fine structure spectroscopy (see Section 11.3), which could, in principle, monitor catalyst operation even in the presence of oil and coal.

A number of ion spectroscopic techniques have already demonstrated their applicability to catalytic systems in the last few years (see Section 11.2.2). They rely on collisions of ions with the surface to determine surface properties and can detect the presence of hydrogen of particular importance in the identification of the nature of carbonaceous materials that lead to the deactivation of coal conversion catalysts.

15.4 SUMMARY AND CONCLUSIONS

In the upgrading of solvent refined coal one of the key points is the search for selective HDN catalysts that avoid or minimize hydrodesulfurization. As there is yet no example for such catalysts, their development would be a long-term research goal while, for the short term, the objective would be optimizing current catalysts and maximizing their HDN activity. Utilization of catalyst–support interactions will be helpful in this respect and also in the stabilization of the catalyst surface area, particularly during regeneration.

Lower H_2S concentrations in SRC upgrading extend the choice of potential catalysts, and many complex oxides, borides, carbides, and nitrides are interesting candidates. New synthesis methods may make it possible to prepare some of these materials with the high surface areas needed for catalytic applications.

The determination of the specific surface area of the catalyst is an important short-term need not only for the SRC-upgrading program but also for other areas for coal liquefaction. It will lead to more reliable control of catalyst performance and deactivation and will permit more accurate evaluation of the activity of catalyst preparations and formulations and of their effectiveness on regeneration.

Of equal importance is the identification of the role of pore-size distribution and surface acidity on the catalyst activity. This work should be directed specifically to SRC, since the composition of the liquids is the determining parameter for the optimization of these properties of the catalyst.

There is also a need for such supporting information as the better understanding of the mechanism of hydrodenitrogenation; of the effect of catalyst morphology, particle size, and composition; and of the thermodynamics of the compounds of interest in the environment expected during SRC upgrading.

REFERENCES

1. Upgrading of Coal Liquids for Use as Power Generation Fuels, Mobil Research and Development Corp., EPRI Rep. 361-1 (1966).
2. Yakabayashi, E., *Jpn. Pet. Inst. J.* **16**, 651 (1973).
3. Schuit, G. C., and Gates, B. C., *AIChE J.* **19**, 417 (1973).

4. McColgan, E. C., and Parsons, B. I., Canadian Mines Branch Research Rep. R273, 1974.
5. Ciapetta, F. G., Dobres, R. M., and Baker, R. W., *in* "Catalysis" (P. H. Emmett, ed.), Vol. VI, p. 495. Von Nostrand–Reinhold, Princeton, New Jersey, 1958.
6. Dillimore, D., Galwey, A., and Rickett, G., *J. Chim. Phys. Phys.-Chim. Biol.* **72**, 1059 (1975).
7. Gauguin, R., Graulier, M., and Papee, D., *in* "Catalysts for the Control of Automotive Pollutants" (J. E. McEvoy, ed.), p. 147. American Chemical Society, Washington, D.C., 1975.
8. Tauster, S. J., *J. Catal.* **26**, 487 (1972).
9. Boudart, M., Cusumano, J. A., and Levy, R. B., New Catalytic Materials for the Liquefaction of Coal, Electric Power Research Institute, Rep. No. RP-415-1, October 30, 1975.
10. Kittrell, J. R., U.S. Patent No. 3,536,605 (1970).
11. Wilhelm, F. C., Climax Molybdenum Rep. L-287-42 (1976).
12. Barz, H., *Mater. Res. Bull.* **8**, 983 (1973).
13. Koros, R. M., Bank, S., Hofmann, J. E., and Kay, M. I., *Am. Chem. Soc. Div. Pet. Chem. Prepr.* **12**(4), B-165 (1967).
14. Voorhoeve, R. J. H., and Stuiver, J. C. M., *J. Catal.* **23**, 228 (1971).
15. Takeuchi, A., Tanaka, K., Toyoshima, I., and Miyahara, K., *J. Catal.* **40**, 94 (1975).
16. Spicer, W. E., Yu, K. Y., Pianetta, P., Lindau, I., and Collins, D., *in* "Surface and Defect Properties of Solids" (J. M. Thompson and M. W. Roberts, eds.), Vol. 5. Univ. of Bradford, The Chemical Society of London, 1972.
17. Ratnasamy, P., *J. Catal.* **40**, 137 (1975).

Chapter 16

Catalytic Hydrocracking of Coal or Lignite to Low Sulfur, Low Nitrogen Liquid Boiler Fuels

The objective of coal liquefaction to boiler fuels is to increase the hydrogen to carbon ratio of the coal and to reduce the concentrations of oxygen, sulfur, nitrogen, and mineral matter. A typical boiler fuel has a H to C atomic ratio around 1.6, and O, S, N, and mineral contents each below 0.5%. The composition of coal, on the other hand, varies considerably with rank, and liquefaction requirements vary accordingly. As shown in Table 16-1, bituminous coal has the highest sulfur concentration of the common coals, lignite has the highest mineral content, and both these coals have high oxygen levels.

Increasing the hydrogen to carbon ratio and removal of impurities require a considerable amount of costly hydrogen. Therefore, minimization of hydrogen

TABLE 16-1

TYPICAL COMPOSITIONS OF COAL[a]

Coal	C	H	O	S	N	Other	H/C atom ratio
Bituminous	73	5	9	3	1.5	8.5	0.82
Subbituminous	71	5	16	0.5	1.5	6.0	0.85
Lignite	64	4.6	18	0.5	1.5	11.4	0.86

[a] In wt%.

consumption is a primary requirement for any catalytic liquefaction process. To do this most effectively, it is necessary to restrict hydrotreating to the removal of the S and N compounds, and the cracking primarily to the center-ring type to maximize liquid yields and minimize the formation of light gases.

16.1 SUMMARY OF MAJOR CONSTRAINTS AND OBJECTIVES

Presently, two general approaches to direct catalytic coal liquefaction are being considered. One uses a heterogeneous catalyst, namely cobalt and molybdenum oxide on alumina, either run in a fixed bed (e.g., the Synthoil process) or fluidized in an ebullient bed, as practiced in the H-Coal process. In the fixed-bed system, coal liquids are contacted with the catalyst either in the presence of coal or in a separate reactor with a hydrogen-donor solvent. The donor solvent is usually hydrogenated and recycled. The other employs a metal halide melt as the catalyst which is $ZnCl_2$ in the Consol process. This system provides improved catalyst–coal contacting and operates at a lower temperature and pressure than the cobalt molybdate system. It has several major drawbacks: high catalyst–coal ratios (up to 3); catalyst deactivation by sulfur, nitrogen, ash, and carbonaceous residues; stringent economic regeneration requirements; and corrosion by the melt.

Catalyst improvements in the mentioned liquefaction processes must overcome a number of constraints. These include:

(1) catalyst deactivation due to the deposition of carbon, metals, and minerals;
(2) catalyst poisoning by nitrogen and sulfur compounds;
(3) catalyst deactivation by sulfidation in the H_2S environment of all liquefaction processes;
(4) limitations in the effectiveness of the catalyst–coal liquid–hydrogen contacting and in diffusion;
(5) high hydrogen consumption and decreased liquid yields due to nonselective cracking;
(6) catalyst and support sintering, particularly during regeneration;
(7) poor heat and mass transfer properties because of improper control of pore-size distribution; and
(8) mechanical degradation with continued use and regeneration.

16.2 IMPACTING AREAS OF RESEARCH DEVELOPMENTS

16.2.1 Short-Term Developments

One of the major obstacles in the evaluation of the activity of existing liquefaction catalysts is the lack of catalyst characterization. To date, catalyst activity data in this area have been analyzed only on the basis of the total catalyst weight or surface area. The latter includes the surface area of the support, and is not necessarily proportional to the total active catalytic surface. Therefore, such information can often be misleading. In recent years a number of interesting leads have been uncovered which may provide a basis for the development of routine techniques for the measurement of nonmetallic surface areas. They include the use of gases such as NO for the characterization of a number of oxides (see Section 4.3.2) and the use of room temperature H_2S and H_2 adsorption for the characterization of Co–Mo/Al_2O_3 catalysts [1]. Further development of such techniques is important for the improvement of the activity of existing HDS and liquefaction catalysts.

For the stabilization of composition and of surface area the use of catalyst–support interaction is an important technique, especially for oxide catalysts and supports (see Section 3.1.2). The inorganic chemistry of compounds such as cobalt and molybdenum oxides or oxysulfides may provide leads for modifications of the support designed to maximize and maintain the active surface area. The formation of complex oxides such as the transition metal–alkaline earth metal oxides are of particular interest in this context [2].

Considerable progress has been made in understanding the role of the support in catalysis. Thus, the acidic properties of the support are important for determining the activity and selectivity of HDS and HDN catalysts [3, p. 84ff]. The control of these acidic properties by the preparation of mixed oxides and the variation of oxide composition offers the opportunity to optimize this property for a given catalytic process. This may be especially important for HDN, where there are indications that rupture of the N—C bond is rate limiting. Therefore, the strength of interaction of the basic N atom with the surface may ultimately determine steady-state catalyst activity (see Section 7.3.2).

The deposition of metals and minerals on the catalyst surface is another problem related to overall catalyst activity maintenance. It is encountered also in petroleum processing, and there has been increasing concern over the last few years about the effect of V and Na on the deactivation and regeneration of HDS catalysts. It is not clear to date whether the effects are physical or chemical. In the case of deactivation, metal accumulation with subsequent blockage of pores may lead to activity loss even without a chemical interaction between the catalyst and the metal impurities. The problem is likely to be especially severe in coal conversion due to the presence of Ti (see Section 6.3.2), the concentration of which can be as high as 0.06% [4]. Using typical throughputs and catalyst loadings for a Synthoil pilot plant [5], for example, this concentration is equivalent to a maximum accumulation of 2% of Ti on the catalyst per day.

The chemical effect of metals and minerals is most dramatically felt during regeneration. Experiments with petroleum HDS catalysts suggest that only catalysts exposed to low levels of minerals and metals can be oxidatively regenerated without irreversible loss in support surface area (see Section 6.3.2). The behavior of coal conversion catalysts has to be studied in more detail to identify possible sources of degradation and to develop means to avoid it. In the preliminary work on the Synthoil process, partially solubilized coal is fed over a catalyst at high linear velocities, giving rise to some attrition of the catalyst surface, and reportedly preventing the accumulation of carbon and mineral matter. It is likely that most cobalt molybdate catalysts will undergo uncontrolled attrition with extensive use. Recent developments in the area of catalyst preparation indicate that a number of additives may be available to reduce high attrition rates [6].

Another problem of regeneration that requires attention is the need to reactivate the catalyst after it has been oxidized during the removal of carbonaceous residues. This places a limitation on the choice of catalytic materials, and a severe constraint on the structural integrity of the catalyst during repeated regeneration–reactivation cycles. The concept of catalyst–support interactions (see Section 3.1.2) is pertinent to this problem. Some of the recently developed applications of this concept to NO_x reduction catalysts illustrate the use of catalyst–support interactions in various chemical environments.

A problem common to all coal liquefaction programs is the choice of test reactor system. This applies to the determination of activity, selectivity, and activity maintenance. Among the autoclave reactors (see Section 8.2.4) only the low thermal inertia reactor can be heated and cooled sufficiently rapidly for effective use in catalyst screening. The most versatile testing system is a three-phase continuous reactor of which several modifications have been developed over the last few years (see Section 8.2.2).

The development of characterization techniques for measuring the active catalyst surface area may find utility for determining the extent of sintering and of chemical changes of the active catalytic surface.

The HDS and HDN catalysts face the problem of support sintering which can be particularly severe during regeneration, especially in the presence of minerals and metals (see Section 6.3.2). Attempts to stabilize supports include cationic doping, which has a pronounced effect on the thermal stability of Al_2O_3 [7]. Such methods should be explored in connection with the stabilization of HDS and HDN catalysts.

The study of sintering phenomena has not yet been extended to HDS and HDN catalysts, partly because of the lack of thorough characterization. It is likely to become an important area in view of the problems that are specific to these systems. The fluxing effect of V_2O_5 on HDS catalysts used in refining of residuum is a case in point. An additional problem is the V_2O_5-catalyzed formation of SO_3 during oxidative regeneration, which leads to sulfating of the support, with subsequent phase change, surface area loss, and mechanical degradation. These reactions are unlikely to occur in the presence of TiO_2 due to the high stability of the sulfate [8,9]. However, the possible effect of other mineral matter in coal should be considered.

16.2.2 Long-Term Developments

Eventually, new catalysts will have to be found to overcome the activity, selectivity (hydrogen consumption and liquid yield), and stability limitations of current processes. A number of developments that are likely to impact in this area are examined in this section.

The properties of the support that play an important role in the behavior of the catalyst include pore-size distribution and acidity. Gel procedures for the synthesis of supports with controlled surface acidity (see Section 5.2.2) provide means for systematically varying this important property of liquefaction catalysts. Similarly, novel preparations of supports with controlled crystallite size and pore-size distribution (see Sections 5.2.2 and 5.4.3) are going to be significant in coal conversion. The innovative procedures for the compounding of mixed oxides with ultrastable zeolites reported in the last few years provide additional compositions of interest since there are indications that these coprecipitated systems have very high HDS and HDN activity [10].

The presence of high H_2S concentrations in coal liquefaction places a stringent constraint on potential catalysts [3]. The few groups that are most promising include oxides, oxysulfides, and sulfides.

A number of complex oxides and sulfides recently synthesized and characterized exhibit interesting structural characteristics such as the metal–metal clusters of molybdenum in $Al_{0.5}Mo_2S_4$ [11] and $Mg_2Mo_3O_8$ [12]. Some of these are available with very high surface areas (see Section 5.2.2) and are suitable for catalytic studies. The possibility of stabilizing valence states such as Mo^{4+} in a complex oxide matrix, and the possible change in oxidation state due to Mo–Mo cluster formation, may have significant effects on the catalytic properties of these materials. Oxysulfide formation in these systems may be particularly important for liquefaction, especially in view of increased activity of HDS catalysts in the presence of oxygen [13].

The effect of oxygen on activity is very different in the case of HDN [14], as is the mechanism and the most effective catalyst. The optimum Co/Mo ratio, for example, is different for HDS and HDN (see Section 7.3.1). New catalysts have to be developed for HDN which can survive the high H_2S environment of the process. A sequential process with two catalyst beds, one optimized for HDS and one for HDN, should be considered. Similarly, a multifunction system incorporating the two optimum catalysts should be designed and tested.

One of the important problems of coal liquefaction is coal–catalyst–hydrogen contacting. In most processes the coal is dissolved by a hydrogen-donating liquid which in turn is rehydrogenated at the catalyst surface. Recent observations suggest that the nature of the coal dissolution is crucial in minimizing hydrogen consumption [15], and that the coal structure is not as aromatic in nature as one would deduce from the analysis of conventionally produced coal liquids. Coal structures consist of significant numbers of one-, two-, and three-ring aromatic and naphthenic units interconnected by paraffinic methylene groups which are believed to participate in condensation reactions during the usual liquefaction processes. Thus, much of the asphaltenic nature of coal liquids may be generated during dissolution by polymerization and condensation reactions.

A technique that may improve coal dissolution is the use of supercritical conditions. While only limited work has been done in this area, there is evidence that such conditions improve coal extraction [16] and considerably increase hydrogen solubility, both factors of importance in coal liquefaction.

Another important aspect of characterization is the use of spectroscopic techniques for surface analysis. A number of promising methods have been studied in the last few years (see Section 11.2), but a longer-range development effort will be required to make them available for routine application. Among the most useful methods are Auger electron spectroscopy, x-ray photoelectron spectroscopy (see Section 11.2.1), ion scattering spectroscopy, secondary ion mass spectroscopy (see Section 11.2.2), and the recently developed x-ray absorption fine structure spectroscopy (see Section 11.3). The latter method promises to provide unique information about the local structure and environment of surface atoms in highly dispersed systems. One particular advantage of ion spectroscopies is their ability to detect hydrogen, an important constituent of surface residues on liquefaction catalysts.

Catalyst improvements often benefit directly from an understanding of the reaction mechanism involved. Reactions of importance to coal liquefaction are HDS, HDN, and polynuclear aromatic cracking. While HDS has received considerable attention in the last few years (see Section 7.4) this is not the case for HDN (see Section 7.3). These two reactions have considerably different rate-determining steps, and it is therefore not surprising that the best catalysts for both reactions are not the same. More studies are needed to relate the difference in mechanisms to the difference in catalyst requirements. Studies that are particularly interesting in this respect involve the effects of simple molecules such as O_2, H_2S, CO, CO_2, and NH_3 on the activity and selectivity of HDN and HDS reactions.

Preliminary results suggest that oxygen plays a more important role in HDS than in HDN [13,14]. The formation and behavior of oxysulfides may therefore be of great importance to the former reaction. The thermodynamics of the formation of oxysulfides, and of their stability at high H_2S concentrations, and in repeated reaction–regeneration cycles should be studied. Similarly, the balance between increased sulfiding resistance and the changes in HDS/HDN activity of catalysts should be examined.

In the case of certain sulfide catalysts there are indications that the edges of particles, and thus particle size, may have a profound effect on the activity [17]. This conclusion follows from a systematic study of the hydrogenation activity and physicochemical properties of the nickel–tungsten sulfide system. The study should be extended to other systems and applied to HDS or HDN reactions. It does point to an area that may be important in the optimization of catalysts for these processes.

In considering the potential use of novel inorganic compounds as catalysts, it becomes apparent that not much is known about the stability, in particular chemical stability in H_2S, of a number of classes of inorganic compounds. Examples are borides, phosphides, and silicides. There is a definite need for more information on the thermodynamics of these compounds in sulfiding and oxidizing atmospheres. Preliminary examination of the stability of a large number of compounds in H_2S indicated that only few would resist sulfidation. They include borides, phosphides, and silicides of the Group VIII metals (see Section 9.5), certain nitrides [18], and oxides (see Section 9.1). More thermodynamic and kinetic information is needed.

16.3 CONCLUSIONS

A number of developments in catalysis and related areas are likely to play an important role in the improvement of the catalytic liquefaction of coal to boiler fuels. Some of these developments have immediate implications for the improvement of existing systems. Others are going to be important in the search for new catalysts and processes.

Of particular importance in the short-term development are techniques for the measurement of the surface area nonmetallic catalysts. Such characterization techniques have implications for the study and improvement of catalytic activity, for the identification of the role of catalyst and support, and for the elucidation of catalyst sintering during reaction and regeneration. Other areas needing study include optimization of the pore-size distribution of the catalyst to minimize deactivation, modifications of the support to improve attrition resistance, the solution of catalyst regeneration problems, and the improvement of HDN in liquefaction processes. For the latter it is important to perform activity studies with real feeds since competitive adsorption may substantially change the behavior observed with model compounds. Among the long-term developments, the application of novel preparative procedures and catalyst formulations are likely to have the most dramatic impact. New materials of interest include complex oxides, sulfides, and oxysulfides. These developments should be complemented by more fundamental studies of the mechanisms of reactions of interest, especially HDN. These advances will also require more information concerning the thermodynamic properties for many of the catalytic materials in the presence of reactive environments.

REFERENCES

1. Dillimore, D., Galwey, A., and Ricket, G., *J. Chim. Phys. Phys.-Chim. Biol.* **72**, 1059 (1975).
2. Ross, P. N., Jr., and Delgass, W. N., *J. Catal.* **33**, 219 (1974).
3. Boudart, M., Cusumano, J. A., and Levy, R. B., New Catalytic Materials for the Liquefaction of Coal, Electric Power Research Institute, Rep. No. RP-415-1, October 30, 1975.
4. Tingey, G. L., and Morrey, J. R., Coal Structure and Reactivity, Battelle Energy Program Rep., p. 40, 1973.
5. Akhtar, S., Mazzocco, N. J., Weintraub, M., and Yavorsky, P. M., Synthoil Process for Converting Coal to Non-Polluting Fuel Oil, presented at the *Synthetic Fuels From Coal Conf., 4th, Oklahoma State Univ.*, May 6–7, 1974.
6. Gauguin, R., Graulier, M., and Papee, D., Thermally stable carriers, *in* "Catalysts for Control of Automotive Pollutants" (J. E. McEvoy, ed.), p. 147. American Chemical Society, Washington, D.C., 1975.
7. Hindin, S. G., and Pond, G. R., German Offen. 2,458,221 (1975); Hindin, S. G., Pond, G. R., and Dettling, J. C., German Offen. 2,458,122 (1975).
8. Cusumano, J. A9, and Levy, R. B., Evaluation of Reactive Solids for SO_2 Removal During Fluidized-Bed Coal Combustion, Electric Power Research Institute, Rep. No. TPS75-603, p. 11, October 1, 1975.
9. Boreskov, G. K., *Adv. Catal.* **15**, 330 (1964).
10. Kittrell, J. R., U.S. Patent No. 3,536,605 (1970).
11. Barz, H., *Mater. Res. Bull.* **8**, 983 (1973).
12. Tauster, S. J., *J. Catal.* **26**, 487 (1972).
13. Kolboe, S., and Amberg, C. H., *Can. J. Chem.* **44**, 2623 (1966).
14. Goudriaan, F., Gierman, H., and Flugter, J. C., *J. Inst. Pet. London* **59**, 40 (1973).
15. Nature and Origin of Asphaltenes in Processed Coal, Electric Power Research Institute, Program No. RP-410, Mobil Res. and Dev. Co.
16. Harrison, J. S., *J. Am. Chem. Soc. Div. Pet. Chem. Prepr.* **21**, 92 (1976).
17. Voorhoeve, R. J. H., and Stuiver, J. C. M., *J. Catal.* **23**, 243 (1971).
18. Levy, R. B., *in* "Advanced Materials in Catalysis" (J. J. Burton and R. L. Garten, eds.). Academic Press, New York, 1977.

Chapter 17

Water–Gas Shift and Methanation Catalysis in the Synthesis of Substitute Natural Gas

17.1 INTRODUCTION

The production of a high Btu gas for use as substitute natural gas (SNG) involves the catalytic conversion of a H_2–CO mixture, the synthesis gas, to methane. Synthesis gas is derived from gas produced by the gasification of coal with O_2 and H_2O. The composition is determined by both the coal properties and the conditions of gasification, as shown in Table 17-1. The amount of methane produced varies from 0 to 26%, and depends mainly on the extent of the low-temperature devolatilization step to which the coal is subjected prior to high-temperature gasification. The H_2 to CO ratio is generally in the range of 1.2–3, although it may be as low as 0.5 as in the Koppers–Totzek gasifier because of the high gasification temperature. The H_2S level is determined by the concentration of sulfur in the raw coal since very little is removed during most gasification processes. An exception is the CO_2 acceptor process of Consolidation Coal Company in which the dolomite, used as a heat carrier, absorbs much of the H_2S and CO_2. The variation in the H_2S concentration arises from the use of both western lignites with a sulfur content of less than 1% and eastern bituminous coals with sulfur levels up to 4%.

The exit gas from the gasifier is freed from particulate matter. It is then scrubbed clean of H_2S either before or after the ratio of H_2 to CO in it is increased

TABLE 17-1

COMPOSITION OF EXIT GAS FROM VARIOUS GASIFICATION PROCESSES

Process	Gas composition (vol%)							Reference
	CO	H_2	CO_2	H_2O	CH_4	H_2S	H_2/CO	
Lurgi (Lurgi Mineralöltechnik G.m.b.h.)	9.2	20.1	14.7	50.2	4.7	(0.3)	2.2	[a]
Koppers–Totzek (Heinrich Koppers G.m.b.h.)	50.4	33.1	5.6	9.6	0	0.3	0.66	[a]
Winkler (Davy Powergas Inc.)	25.7	32.2	15.8	23.1	2.4	0.3	1.3	[a]
Synthane (US Bureau of Mines)	10.5	17.5	18.2	37.1	15.4	0.3	1.7	[a]
Bi-Gas (Bituminous Coal Research Inc.)	22.9	12.7	7.3	48.0	8.1	0.7	0.55	[a]
CO_2 Acceptor (Consolidation Coal Co.)	14.1	44.6	5.5	17.1	17.3	0.03	3.2	[a]
HYGAS: steam oxygen (Institute of Gas Technology)	18.0	22.8	18.5	24.4	14.1	0.9	1.3	[b]
HYGAS: steam–iron (Institute of Gas Technology)	7.4	22.5	7.1	32.9	26.2	1.5	3.0	[b]

[a] From Bodle and Vyas [1], courtesy of the Petroleum Publishing Company.
[b] From U.S. Energy Research and Development Administration [2].

to a value near 3 by the water–gas shift reaction, since shift catalysts resistant to sulfur are available. On the other hand, methanation catalysts, generally massive Raney Ni or supported Ni, are extremely sensitive to sulfur and require nearly quantitative removal of sulfur down to a level between 0.01 and 0.2 ppm. The contribution of the shift, purification, and methanation steps to the cost of the methane produced is estimated to be in the range of 20% [3]. Only a small part (~ 5%) of the cost is associated with actual catalyst costs. The primary costs are connected with the gasification step, and amount, depending on the process, to between 60 and 80% of the methane value. Process savings can result from two types of improvement of the methanation catalyst: better sulfur tolerance will permit less costly sulfur removal systems, and improved thermal stability will allow operation at higher temperatures and more efficient waste heat recovery.

17.2 MAJOR CATALYST PROBLEMS AND OBJECTIVES

The problems associated with the water–gas shift and methanation reactions will be discussed under three headings: poisoning, thermal stability, and carbon deposition.

17.2.1 Poisoning

The first catalytic process in the conversion of synthesis gas is the water–gas shift reaction. Commercial catalysts for this process include iron oxide–chromium oxide catalysts operating at 315–485°C and more active zinc oxide–copper oxide catalysts [4] operating at 175–350°C. The latter are sensitive to low levels of sulfur, while the former can tolerate H_2S levels of 500–1000 ppm [4]. Low-temperature operation is preferred because it allows higher equilibrium conversion to H_2. For use in hydrocarbon synthesis requiring lower conversions of CO to H_2, iron oxide–chromia catalysts are acceptable. Recently, low-temperature shift catalysts with significantly increased sulfur tolerance have been developed [5,6], so that high levels of sulfur are now not a major problem for the water–gas shift reaction.

The magnitude of the sulfur poisoning problem of methanation catalysts can be seen from the life of a 25% Ni catalyst shown in Table 17-2 and calculated on the following assumptions. The adsorption of sulfur from the reactant stream is quantitative, and the sulfided catalyst is inactive. Deactivation proceeds through the catalyst bed, and the catalyst must be replaced when 50% of the bed is sulfided. At an H_2S level of 10 ppm in the reactant stream, a surface sulfide is formed in five days, and in twenty days completely converts 50% of the Ni to bulk NiS. Only in the range of 0.1–1 ppm of H_2S does the catalyst lifetime become sufficiently long for a practical process.

TABLE 17-2

EXPECTED CATALYST LIFETIME IN SULFUR-CONTAINING STREAMS

Process assumptions	
H_2S concentration	As listed
Gas space velocity (h^{-1})	10,000
Catalyst deactivated at 50% sulfidation	
Catalyst properties	
Ni content (wt%)	25
Apparent bulk density (gm cm^{-3})	1
Ni dispersion[a]	0.25

Calculated catalyst lifetime (days)

	Type of sulfidation	
H_2S concentration (vol/vol)	Bulk NiS (days)	Surface NiS (days)
10 ppm	20	5
1 ppm	200	50
0.1 ppm	2000	500

[a] Ratio of surface Ni to total Ni present in sample.

A level of 1 ppm H_2S in a reactant stream containing 50% H_2 corresponds to an H_2S/H_2 ratio of 2×10^{-6}. At this H_2S/H_2 ratio, bulk nickel sulfide is not stable at the temperature of operation for most methanation processes (300–450°C) [7]. Benard has found that a surface sulfide layer on Ni is stable at much lower H_2S/H_2 ratios [8,9]. Comparative data presented by Rostrup-Nielsen at 550°C show that an H_2S/H_2 ratio of 2×10^{-6} is sufficient to form a saturated sulfur layer on Ni, whereas a ratio of 10^{-3} is required to form the bulk sulfide [10].

Calculated catalyst lifetimes such as those described above can be compared with recent data obtained at Sasol, South Africa, in a pilot plant scale methanation operation [11]. Using a supported catalyst with a high Ni content, an average level of 0.05 ppm H_2S caused a slow but noticeable deactivation of the inlet portion of the catalyst bed, but no deactivation farther into the catalyst bed even after fifty days on stream. On the other hand, at a level of 3 ppm H_2S, a very rapid deactivation of the catalyst occurred much farther into the catalyst bed after only six days of operation. While operation for fifty days with a sulfur level of 0.05 ppm hardly changed the zone of maximum conversion in the bed, twenty days at a level of 3 ppm H_2S moved the maximum conversion zone from 22 to 44% of the bed length.

The form of the Ni catalyst does not appear to have a great effect on its sulfur tolerance or sulfur capacity. Similar rapid sulfur poisoning was observed for Raney Ni catalysts used in fluidized bed operation [12] or as flame-sprayed coatings on tube-wall reactors [13].

The regeneration of sulfur-poisoned Ni catalysts is rather difficult [2, Vol. 3, Part 5, pp. 53,80]. It would require either reductive or oxidative processing. The former would involve the use of large amounts of costly sulfur-free hydrogen or synthesis gas, and may be extremely slow even at temperatures as high as 700°C [10]. Oxidative regeneration would entail treatment in steam (17-1) and (17-2) or calcination in air (17-3):

$$NiS + H_2o \rightarrow NiO + H_2S \quad (17\text{-}1)$$

$$H_2S + 2H_2O \rightarrow SO_2 + 3H_2 \quad (17\text{-}2)$$

$$NiS + {}^{3}/{}_{2}O_2 \rightarrow NiO + SO_2 \quad (17\text{-}3)$$

The steam process requires complete conversion to the metal oxide, but does appear to rapidly remove much of the nickel sulfide while the oxygen treatment leads to sulfation of the catalyst and retention of most of the sulfur even at 750°C [10]. The oxidation step could have an irreversible adverse effect on the massive nickel catalysts because of spinel formation from nickel oxide and the alumina support.

The only other poison of consequence is chlorine. Low-temperature Cu/Zn/Al water–gas shift catalysts are poisoned by the formation of $ZnCl_2$ with resultant sintering of the active copper phase [14]. The detrimental level of chloride is much lower than the tolerable sulfur level, and, after exposure of the catalyst to chloride, deactivation continues slowly as the active component deteriorates. The chloride is usually introduced from the water supply, and can be prevented from entering the primary reactor by using a guard chamber filled with low temperature water–gas shift catalyst.

17.2.2 Thermal Stability

A highly exothermic reaction such as methanation can be carried out in two basically different modes. In one, the reaction temperature is kept low by product gas recycle, by using tube-wall reactors or a liquid heat removal medium. High catalyst thermal stability is not a prime requirement for such systems. Alternatively, the reactor can be operated adiabatically at the maximum temperature obtained, depending on the conversion limit imposed by equilibrium. Under such conditions reaction heat can be recovered in the form of high-temperature, high-pressure steam, improving overall thermal efficiency. A process proposed by R. M. Parsons Company uses a series of methanation reactors operating at successively lower temperatures [15]. The initial reactors with an outlet temperature of 770°C produce 40–50% of the methane. The final reactors convert the residual CO at substantially lower temperatures. This process requires catalysts of extreme thermal stability. The catalyst in the initial reactors may be less active and less tolerant to sulfur poisoning since both bulk and surface sulfides have lower stability at the high operating temperatures.

17.2.3 Carbon Deposition

A third major problem area stems from the Boudouard reaction, the disproportionation of CO to CO_2 and carbon:

$$2CO \rightleftharpoons CO_2 + C \tag{17-4}$$

Carbon monoxide present in the reaction stream can affect the equilibrium of reaction (17-4) through the water–gas shift (17-5) or the methanation reactions (17-6):

$$H_2O + CO \leftrightarrows H_2 + CO_2 \tag{17-5}$$

$$3H_2 + CO \leftrightarrows H_2O + CH_4 \tag{17-6}$$

If it is assumed that these reactions are in equilibrium, the extent of the production of elemental carbon can be readily calculated for any gas composition from the known free energies of these reactions [16]. A particularly useful presentation format is a ternary diagram described by Haynes *et al.* [17], showing the gas compositions in which carbon can form. For example, at 2 atm a stoichiometric mixture of $3H_2$ + CO will not deposit carbon below a temperature of 430°C. The presence of water further reduces the extent of carbon formation. Actually, more often than not, equilibrium is not established because of kinetic limitations, and therefore these thermodynamic data should be applied with caution. Thus in bench scale tests of Ni catalysts it is found that no carbon deposition occurs under thermodynamically favorable conditions [17, 18] unless some iron is present. In general, carbon deposition is hindered at high H_2/CO ratios and in the presence of water vapor. Increased H_2 partial pressures shift the equilibrium away from carbon, while water acts through the water–gas shift reaction to produce H_2 at the surface or by gasification of the deposited carbon.

In actual use, Ni catalysts operating at temperatures up to 450°C and with 20–50 mol% H_2O in the reactant stream maintain activity for prolonged periods at H_2/CO ratios from 5.8 to 2.0 [11]. Catalysts deactivated by carbon deposition reportedly cannot be regenerated [2, Vol. 3, Part V, pp. 53, 80]. However, not enough systematic work has been done on this type of regeneration to exclude its feasibility completely.

Carbon deposition is generally not a problem in water–gas shift catalysis because of the presence of excess water which inhibits carbon deposition, maintains the iron in iron–chromium catalysts in the oxidized state, and makes it inactive for the Boudouard reaction.

17.3 IMPACTING AREAS AND RESEARCH DEVELOPMENTS

The problems described in the preceding sections point to a number of desirable catalyst properties that require development. For methanation catalysts these are sulfur tolerance to a level of 10–1000 ppm of H_2S and thermal stability at 500–800°C. Minimal carbon deposition is desirable for both methanation and shift catalysts. The following two sections present approaches to poison-resistant and thermally stable catalysts based on advances in catalysis and related disciplines. The next two sections describe some engineering features for controlling the temperature in methanation, and the potential of homogeneous catalysts. The final section is devoted to desirable areas of supporting research covering the mechanisms of methanation and sulfur poisoning, and the nature of the catalyst surface.

17.3.1 Poison-Resistant Catalysts

In the search for a catalyst tolerant to the presence of 10–1000 ppm H_2S in the reactant stream, it is necessary to use the proper test procedure. In the past, the addition of H_2S to integral reactors operating at high conversion gave data indicative of sulfur capacity of the catalyst, since after some arbitrary loss of catalytic activity the test was terminated, and the catalyst noted as poisoned. To assess tolerance to sulfur, the activity must be measured after the catalyst has come to steady state with the H_2S level in the reactant stream. An effective procedure has been described by Dalla Betta *et al.* [19,20] in which the activity of a catalyst saturated with sulfur in an environment of 10 ppm or more H_2S is measured.

At 10–1000 ppm of H_2S bulk sulfidation does not occur on Ni at the temperature generally used in methanation so that sulfur poisoning is the result of an inactive surface sulfide layer. The problem becomes one of preventing its formation. According to sketchy data, cluster or alloy formation can reduce poisoning effects [21]. The requirement is basically that the free energy of formation of the alloy be sufficiently large to inhibit the formation of the surface sulfide from H_2S. The study of Dalla Betta *et al.* [19] shows surprisingly high methanation activity of Ni on certain supports in the presence of 10 ppm H_2S. The cause is not well established, but may be associated with a surface layer containing carbon that inhibits the formation of the inactive sulfide [20]. A carbon layer on Ni is not sufficiently stable to exclude sulfide formation completely since the nature of the support and the presence of water in the reactant stream cause wide variations in the observed sulfur resistance. The striking effect of the support suggests that metal–support interactions (see Chapter 3) may be an area worthy of investigation. The energy of interaction of the metal with the support can be adjusted by varying the support [22] and metal particle size. Decreasing particle size also increases the effect of promotors present in the support. Techniques for the preparation of highly dispersed metals are fairly well developed (see Section 5.2). Work along this line should cover activity measurements in the presence of H_2S at conditions as close to actual operating conditions as possible in order to allow for effects of reaction temperature and of water vapor.

Another study that may yield benefits is the catalytic testing of intermetallic compounds such as $ZrPt_3$ and some of the more common carbides, nitrides, and borides in the presence of 10–1000 ppm of H_2S. Nickel boride has a hydrogenation activity similar to that of Ni [23]. Although these materials are not expected to be stable at 1% H_2S (see Sections 9.4 and 9.5; Boudart *et al.* [24]) they may have sufficient stability at lower sulfur levels.

Materials resistant to formation of bulk sulfide in the presence of 10–1000 ppm of H_2S could still be poisoned by surface sulfidation. This mode of poisoning has not yet been established. Neither is it known whether partial sulfidation of the surface will totally deactivate the metal. The development of catalysts totally resistant to sulfur requires long-range studies in several areas. One would cover the surface chemistry of sulfide formation in which the electron spectroscopy techniques could prove useful.

A second area would concern strong interactions involving the active metal, e.g., Ni, and the support or another material as a means for improving its sulfur resistance. If this interaction results in compound formation, loss in activity could be expected but such loss could be made up by high sulfur tolerance and operation at higher temperature. One interesting interaction is the withdrawal of electrons from platinum metals by zeolite supports, resulting in improved sulfur resistance. Application of this effect to methanation catalysts is worth studying. A third area would be devoted to the evaluation of various compounds discussed in Chapters 9 and 10, since so far few materials other than transition metals have been tested for methanation. Interestingly, even the platinum metals may deserve further examination, in spite of their low activity, in view of their sulfur resistance and high temperature stability [25].

Carbon deposition could be avoided by the selection of proper components for alloy catalysts to give increased hydrogenation activity of the surface and to decrease its tendency to form unsaturated polymeric residues [26]. They are thought to be coke precursors and to require at least three contiguous metal sites for formation [27].

17.3.2 Thermally Stable Catalysts

Catalysts capable of operation at higher temperatures can be developed either by stabilizing supported Ni, or by evaluating new catalytic materials possessing high thermal stability. In both approaches some work will involve technology already tested or partially developed for related systems and reactions.

Methanation is carried out in a highly reducing atmosphere in which Ni is in the reduced state and its interaction with the support is small. As the maximum operation temperature is raised from 450 to 700–800°C, severe sintering is expected to occur. Addition of structural promotors, such as chromia [23], may help in stabilizing the Ni surface area. For providing high surface area supports, one could consider using ultrastable zeolites which withstand temperatures up to 800°C in atmospheres containing water vapor [24, p.111].

Alloy and cluster formation may offer increased thermal stability. Ruthenium decreases the sintering tendency of copper in RuCu clusters compared to that of pure supported copper [28]. Similar behavior has been noted in other alloy systems [29], and may be used to increase the thermal stability of Ni or, alternatively, to construct other active systems with high thermal stability.

17.3.3 Engineering Features

The highly exothermic nature of methanation and its equilibrium limitation at elevated temperatures has motivated much process research and has resulted in the application of unusual engineering concepts. Thus, hot and cold product gas recycle was investigated as a technique to limit the maximum adiabatic reaction temperature in both fixed-bed and fluid-bed reactors. The recycling of large volumes of gas caused a substantial pressure drop through the catalyst bed and led to the use of tube-wall methanation reactors. The catalytic material employed was Raney Ni alloy flame-sprayed onto the tube walls. These catalysts show very good activity but suffer from the problems of other Ni catalysts: poor thermal stability and low sulfur tolerance [30]. The thermal stability is aided somewhat by the very good heat transfer properties of the thin Ni coating on the heat exchanger surface. The tube-wall reactor places a limitation on the operating temperature and will show temperature gradients and thermal degradation at very high throughputs. The development of thermally stable and sulfur-tolerant materials would make the tube-wall reactor more useful. For this purpose composite materials should be considered and techniques developed for applying them to the reactor wall.

Another approach to controlling the temperature of the methanation is the use of a countercurrent liquid flow through the reactor for removing heat. Such a liquid phase methanation reactor has the advantages of efficiency, simple design, and high throughput, but shows relatively low activity because the presence of a liquid layer limits the access of the reactants to the catalyst surface [31]. Improved catalyst activity and accessibility would permit high throughput at a controlled temperature so that the reaction can continue to completion.

17.3.4 Homogeneous Catalysts

Homogeneous catalysts could be advantageously applied to CO hydrogenation since the reactants and products are gaseous and separation problems would be minimized. The liquid medium would provide efficient heat removal. It has also been suggested that homogeneous catalysts could be sulfur tolerant, although this has not been demonstrated [32]. Muetterties and co-workers found that the osmium and iridium cluster compounds, $Os_3(CO)_{12}$ and $Ir_4(CO)_{12}$, catalyze the hydrogenation of CO to methane [33]. At 140°C and 2 atm, these catalysts selectively produce methane at specific rates which compare favorably to the rates measured by Vannice on supported Ir [25]. It was also observed that ligand substitution greatly affected the reaction rate and the product distribution. The substitution of triphenylphosphine for the carbonyl ligands increased the synthesis rate three times and led to the production of methane, ethane, and propane. An effort was made to prove the homogeneity of the reaction system, but this was not established unambiguously. In view of these encouraging results, further study of methanation by homogeneous catalysts appears warranted.

17.3.5 Supporting Research

The problems associated with methanation catalysis described above indicate that more basic supporting research would be valuable in three areas: the mechanism of CO hydrogenation, the mechanism of sulfur poisoning, and the identification of surface phases of the catalyst present during the reaction.

The mechanism of catalytic CO/H_2 synthesis is not well known (see Section 7.1), but indications are that the reaction proceeds by hydrogenation of a surface enol complex, ═CH(OH), implying the need for good hydrogenation activity. On the other hand, if C—O bond breaking was the rate-controlling step, surface properties such as variable surface valence or the formation of a stable surface carbide would be desirable. Knowledge of the methanation mechanism would facilitate the initial evaluation of catalytic materials.

Recent data [20] show that the activity of a certain supported Ni catalyst decreases only by a factor of 2 in the presence of 10 ppm H_2S. In the presence of water vapor the activity drops precipitously by 3 orders of magnitude, but is restored on removing the water vapor from the reactant stream. An explanation for these observations is not yet available and may require the use of a technique such as extended x-ray absorption edge fine structure spectroscopy (EXAFS) to characterize the bulk and surface phases present during reaction (see Section 11.3). In this sulfur tolerance work [20] it was not ascertained to what extent, if any, the surface was covered by sulfur in the presence of H_2S to explain the high activity observed. If there was an inhibition of the formation of the surface sulfide, its elucidation would be useful in understanding sulfur tolerance.

Characterization of supported catalysts during reaction is a difficult but necessary task in the case of sulfur poisoning. Electron spectroscopy techniques for the study of surface layers require high vacuum to permit energy analysis of the emitted electrons and can not be applied to reacting systems. Techniques employing more penetrating radiation, such as Mössbauer spectroscopy, infrared spectroscopy, and EXAFS, could be used at conventional conditions but cannot distinguish readily between bulk and surface layer unless the catalytic phase is in a highly dispersed form. In situ studies of reacting systems and the development of appropriate physical techniques should be pursued.

17.4 CONCLUSIONS

The two primary problems encountered with methanation and water–gas shift catalysts are sulfur poisoning and thermal instability. In order to ensure good activity maintenance, the catalyst is not necessarily operated under the most efficient and most economical conditions. Thus, for conventional nickel methanation catalysts the synthesis gas feed must contain less than 0.1 ppm of H_2S. This constraint requires cooling the gas from the shift reactor, removal of the S by cold scrubbing, and reheating it before it enters the methanator. It would be more economical to use a hot scrubbing process, but that would leave 10–1000 ppm sulfur in the feed gas to the methanator and would require sulfur-tolerant catalysts.

Some progress has been made in the development of sulfur-tolerant systems. Catalyst–support interactions are known to alter the chemical properties of the catalytic phase and can reduce its sensitivity to sulfur. One example is the decreased sulfur sensitivity of nickel on ZrO_2 [20] as compared with that of nickel on Al_2O_3. The preparation of composite highly dispersed materials by new techniques will be useful in exploring and extending the catalyst–support interaction concept for sulfur tolerance. At low levels of sulfur (< 100 ppm), alloys and the intermetallic compounds developed during the past decade may find application in improving poison resistance. The vast area of new catalytic materials from inorganic chemistry could also have a favorable impact. Many metal cluster oxides such as $Mg_2Mo_3O_8$ have shown promise, but have not been studied for methanation or water–gas shift catalysis.

In the short term, experimental leads based on the use of catalyst–support effects, alloys, and other new multicomponent materials should be exploited in the development of catalysts resistant to sulfur and to high temperatures. In the long term, a concerted effort will be required to clarify the mode of interaction of catalysts with sulfur and the nature of their thermal degradation.

Ultimately, the development of efficient economic methanation and water–gas shift processes will depend upon the appropriate combination of catalytic activity and activity maintenance. The latter is influenced by both sulfur poisoning and thermal degradation. The contribution of both is strongly dependent upon the mode of process operation. In the usual low-temperature operation, activity maintenance is likely to be affected more by poisoning effects than by thermal degradation. On the other hand, thermal stability can be more important than sulfur tolerance for high-temperature operation, which requires no gas recycle and which leads to greater thermal efficiency. Recent work [16] suggests that this latter mode may be the more cost effective for future processes. If this is the case, increased emphasis should be put on extending catalyst activity maintenance by inhibition of thermal degradation.

REFERENCES

1. Bodle, W. W., and Vyas, K. C., *Oil Gas J.* 73 (Aug. 26, 1974).
2. U.S. Energy Research and Development Administration, Hy-Gas: 1964 to 1972, Pipeline Gas from Coal, Research and Development Rep. No. 22 (Final), Vol. 1, p.1–11, September 1972.
3. Mills, G. A., *Environ. Sci. Tech.* **5**, 1178 (1971).
4. Thomas, C. L., "Catalytic Processes and Proven Catalysts," p. 104f. Academic Press, New York, 1970.
5. Aldridge, C. L., U.S. Patent No. 3,755,556 (1973).
6. Aldridge, C. L., U.S. Patent No. 3,615,216 (1971).
7. Rosenqvist, T., *J. Iron Steel Inst.* **176**, 37 (1954).
8. Benard, J., *Catal. Rev.* **3** (1), 93 (1969).

9. Perdereau, M., and Oudar, J., *Surface Sci.* **20**, 80 (1970).
10. Rostrup-Nielsen, J. R., *J. Catal.* **21**, 171 (1971).
11. Eisenlohr, E. H., Moeller, F. W., and Dry, M., *Am. Chem. Soc. Div. Fuel Chem. Prepr.* **19** (3), 1 (1974); "Methanation of Synthesis Gas" (L. Seglin, ed.), p. 113. American Chemical Society, Washington, D.C., 1975.
12. Dirkson, H. A., and Linden, H. R., *Inst. Gas Technol. Res. Bull.* No. 31 (1963).
13. Demeter, J. J., Youngblood, A. J., Field, J. H., and Bienstock, D., U.S. Bureau of Mines Rep. of Inv. 7033 (1967).
14. Young, P. W., and Clark, C. B., *Chem. Eng. Progr.* **69** (5), 69 (1973).
15. White, G. A., Roszkowski, T. R., and Stanbridge, D. W., *Am. Chem. Soc. Div. Fuel Chem. Prepr.* **19** (3), 57 (1974); "Methanation of Synthesis Gas" (L. Seglin, ed.), p. 138. American Chemical Society, Washington, D.C., 1975.
16. Anderson, R. B., *in* "Catalysis" (P. H. Emmett, ed.), Vol. IV, p. 1. Van Nostrand–Reinhold, Princeton, New Jersey, 1961.
17. Haynes, W. P., Elliott, J. J., and Forney, A. J., *Am. Chem. Soc. Div. Fuel Chem. Prepr.* **16** (2), 47 (1972).
18. Haynes, W. P., Forney, A. J., Elliott, J. J., and Pennline, N. W., *Am. Chem. Soc. Div. Fuel Chem. Prepr.* **19** (3) (1974); "Methanation of Synthesis Gas" (L. Seglin, ed.), p. 87. American Chemical Society, Washington, D.C., 1975.
19. Dalla Betta, R. A., Shelef, M., and Piken, A. G., *J. Catal.* **40**, 173 (1975).
20. Dalla Betta, R. A., and Shelef, M., *Am. Chem. Soc. Div. Fuel Chem. Prepr.* **21** (4), 43 (1976).
21. Bartholomew, C. H., Quarterly Tech. Progr. Rep. (April 22–July 22, 1975), ERDA Contract No. E(49-18)-1790, August 6, 1975.
22. Geus, J. W., *Int. Symp. Sci. Basis Catal. Preparat., Brussels* October 14–17, 1975.
23. Mears, D. E., and Boudart, M., *AIChE J.* **12**, 313 (1966).
24. Boudart, M., Cusumano, J. A., and Levy, R. B., New Catalytic Materials for the Liquefaction of Coal, sponsored by the Electric Power Research Institute, Palo Alto, California Rep. No. RP-415-1, October 30, 1975.
25. Vannice, M. A., *J. Catal.* **37,** 449 (1975).
26. Sinfelt, J. H., Carter, J. L., and Yates, D. J. C., *J. Catal.* **24**, 283 (1972).
27. Clarke, J. K. A., *Chem. Rev.* **75**, 291 (1975).
28. Sinfelt, J. H., *J. Catal.* **29**, 308 (1973).
29. Sinfelt, J. H., *J. Catal.* **29**, 308 (1973); Myers, J. W., and Prange, F. A., U.S. Patent No. 2,911,357 (1959).
30. Field, J. H., Demeter, J. J., Forney, A. J., and Bienstock, D., *Ind. Eng. Chem. Prod. Des. Dev.* **3**, 150 (1964).
31. U.S. Energy Research and Development Administration Liquid Phase Methanation, Rep. No. 78, 1974; Blum, D. B., Sherwin, M. B., and Frank, M. E., "Methanation of Synthesis Gas" (L. Seglin, ed.), p. 149. American Chemical Society, Washington, D.C., 1975.
32. NSF Workshop on Fundamental Research in Homogeneous Catalysis as Related to U.S. Energy Problems, held at Stanford Univ. (December 4–6, 1974).
33. Thomas, M. G., Beier, B. F., and Muetterties, E. L., *J. Am. Chem. Soc.* **98**, 1296 (1976).

Chapter 18

Catalytic Gasification of Coal or Lignite to Synthesis Gas and Substitute Natural Gas

Among the various alternative coal conversion techniques, gasification is in the most advanced stage. The gasification processes of Lurgi, Winkler, and Koppers–Totzek are in commercial use, while those of Hygas, Bigas, CO_2 Acceptor, Synthane, and Kellog are in the development stage. The objective of these processes is to convert coal to gas for use as fuel for power generation or as feed for upgrading operations.

In the most common approach (see Section 7.6), coal is reacted with oxygen and steam to give a mixture of carbon monoxide, hydrogen, methane, and products of combustion. Alternate routes include reaction with air and steam to produce a low Btu gas (~150–200 Btu ft^{-3}) and reaction with hydrogen (hydrogasification) to maximize the production of methane.

In steam gasification (18-1) the reaction of carbon

$$C + H_2O \rightleftharpoons CO + H_2 \tag{18-1}$$

with steam is highly endothermic. This places a considerable economic burden on the process. In most cases, the heat requirement is provided by the oxidation of part of the coal (18-2):

$$C + \tfrac{1}{2}O_2 \rightleftharpoons CO \tag{18-2}$$

This often amounts to one third of the energy in the coal. Operation at lower temperatures is therefore desirable, and provides an incentive for the development of catalytic steam gasification processes. Lower temperatures also favor CH_4 formation, improve thermal efficiency, and increase the Btu content of the product gas.

In the following, the thermodynamic and catalytic constraints of gasification will be discussed first, followed by a review of some concepts and ideas relevant to the development of improved catalytic gasification systems.

18.1 THERMODYNAMIC AND CATALYTIC CONSTRAINTS

The gasification of coal has thermodynamic limitations, as shown by Fig. 18-1, representing the standard free energy changes of the steam gasification reaction (18-1) and of the following reactions of carbon and carbon monoxide:

$$C + 2H_2 \rightleftharpoons CH_4 \quad (18\text{-}3)$$

$$C + CO_2 \rightleftharpoons 2CO \quad (18\text{-}4)$$

$$CO + H_2O \rightleftharpoons CO_2 + H_2 \quad (18\text{-}5)$$

$$CO + 3H_2 \rightleftharpoons CH_4 + H_2O \quad (18\text{-}6)$$

It will be noted that only the steam gasification (18-1) and the reverse Boudouard reactions (18-4) are favored by high temperatures. By contrast, the methane-producing reactions (18-3) and (18-6) are favored by low temperatures. The hydrogasification reaction (18-3) is very slow in the absence of a catalyst [2], and the catalysts so far tested [3] were severely poisoned by sulfur.

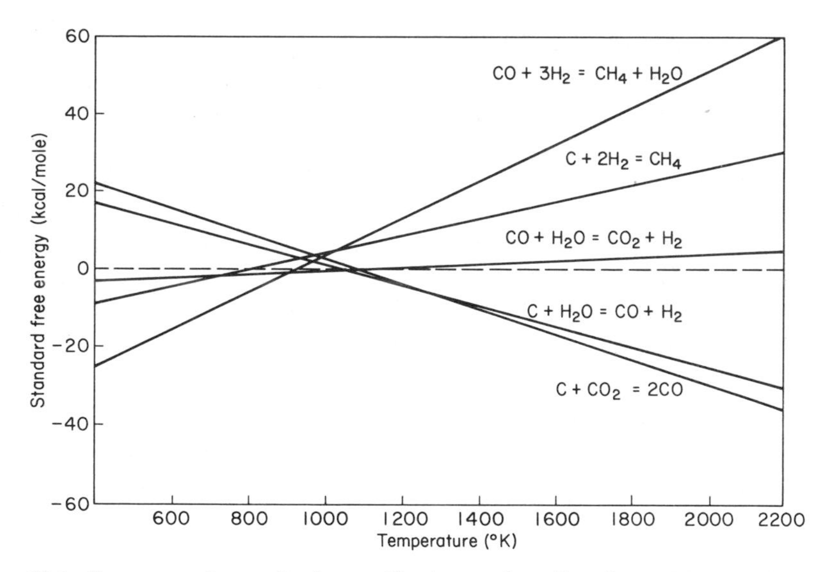

FIG. 18.1 Free energy change of various gasification reactions. From Lowry [1], courtesy of John Wiley and Sons.

Another reaction that contributes to CH_4 formation during gasification is the initial devolatilization of the coal (see Section 7.6). Very little is known about the effect of catalysts on the devolatilization but it appears that acid catalysts such as $SnCl_2$ and $ZnCl_2$ increase the gas yield in the initial coal conversion stages.

Major problems connected with the use of catalysts for the above gasification reactions include the contacting of the catalyst with coal, its recovery, and deactivation.

18.1.1 Catalyst–Coal Contact

The problem of contacting the catalyst with coal prevails in all schemes of coal conversion. It is most severe in gasification since one deals with a gas–solid reaction rather than with a gas–liquid reaction. Current catalytic gasification processes use salts which are in the molten state at reaction temperature. These include $FeCl_2$, $SnCl_2$, $KHCO_3$, and K_2CO_3 [4], among which the latter used in the Kellog process appears to be the most effective. These salts also lower the reaction temperature, and serve as heat-transfer media insuring uniform distribution of heat.

While molten salts are used at a level of 20% [5], significantly higher amounts of metal compounds are required to catalyze the methanation reaction (18-6). Catalyst to coal ratios of 1:1 are not uncommon, making these systems unrealistic from an economic standpoint. The need for such high ratios is due to poor catalyst coal contacting.

Otto and Shelef found that the addition of an aqueous solution of a nickel salt increased the steam gasification rate of char by a factor of five [6]. On the other hand, the catalytic effect was negligible when 2.4 wt% nickel was added in the form of a powder, showing the importance of good contact between catalyst and char. Even impregnation had no catalytic effect when the catalyst was added to the coal before devolatilization [7]. This effect is apparently due to deactivation during the initial pyrolysis step when the coal passes through the plastic phase and either coats the catalyst or poisons it by contact with sulfur and mineral components.

For successful operation, it is likely that coal gasification catalysts will be either in the liquid state, uniformly impregnated on the coal char before reaction, or ideally, well dispersed and highly mobile, thus maintaining contact with the coal during gasification. Because of the continuously changing catalyst–coal interface, even an initially uniform impregnation may prove insufficient for good contacting. Such a dynamic system also complicates the experimental work because of the difficulty in properly defining the catalyst. Recent work with catalytic materials that have some volatility in the gasification environment show a sustained catalytic effect even after a portion of the coal has been reacted [7]. This is probably a direct result of maintaining catalyst–coal contact.

18.1.2 Catalyst Deactivation

There are two modes of catalyst deactivation that are important in gasification: interaction with sulfur, and with mineral matter. Sulfur presents the most severe problem, since few materials will survive the high sulfur concentrations of some coals [8]. Certainly none of the transition metals is likely to remain in the metallic state at these conditions. The best catalysts for methanation (such as Ni and Ru) will therefore be completely impractical in coal gasification, as has been adequately demonstrated by work which showed rapid deactivation [9,10]. The same applies to Ni and Pt catalysts considered for hydrogasification (18-3) [3].

The problem of the interaction of sulfur with such salts as Na_2CO_3 and K_2CO_3 is not as severe as that involving metals. In fact several schemes attempt to take advantage of this interaction to remove some of the sulfur during gasification [11]. At the reducing conditions of gasification, these salts form sulfides which would then have to be regenerated or discarded. So far there has not been much success with the regeneration of such sulfide species. Interest in this problem is related to current work on the fluidized-bed combustion of coal in which dolomite and limestone are used to remove SO_2 during combustion by the formation of $CaSO_4$ [12]. Discarding of the sulfide or sulfate may lead, in the long run, to environmental problems even more objectionable than air pollution by SO_2 [13].

The other deactivation problem is the interaction of the catalyst with the mineral matter in the coal, the composition of which is shown in Table 18-1. Components such as the oxides of Na, K, Fe, Al, and particularly Si, are likely to be of greatest concern because of their abundance. The phase diagrams of the Na_2O–SiO_2 and the K_2O–SiO_2 systems [15] show a number of eutectics melting in the range of 742–846°C. Catalysts containing alkalies would suffer losses by forming eutectics which would be difficult to recover, reprocess, or regenerate.

TABLE 18-1
TYPICAL COMPOSITION OF MINERAL MATTER IN BITUMINOUS COALS[a]

Constituent	Percent of ash
SiO_2	20–60
Al_2O_3	10–35
Fe_2O_3	5–35
CuO	1–20
MgO	0.3–4
TiO_2	0.5–2.5
$Na_2O + K_2O$	1–4

[a] From Ode [14], courtesy of John Wiley and Sons.

The problem may not be as severe in the reducing environment of gasification but would preclude oxidative regeneration of the catalyst if this were required.

18.2 IMPORTANT RESEARCH OBJECTIVES

In view of the constraints discussed, it is expected that only inexpensive, readily available materials will be practical as catalysts for coal gasification. Since they are likely to be used only on a once-through basis, their disposal will have to be environmentally acceptable. In view of the large quantities of coal to be gasified, this requires a low level of added catalytic material or an ash–catalyst combination which is useful or at least unobjectionable.

The following discussion concerns concepts and ideas that are likely to be important in the development of improved catalytic gasification processes. It will cover testing, catalysts, preparative procedures, the mechanism of gasification, and some exploratory programs.

18.2.1 Testing Procedures

One of the challenges of a development program is the choice of a laboratory reactor which provides representative information about the reactions that are important on a commercial scale [6]. A method of particular interest to coal gasification uses the thermogravimetric balance (Section 8.3). The technique, applied in a number of investigations of the reactivity of graphite [16] and the combustion of coal [12] monitors the loss of sample weight at controlled temperature, pressure, and flow conditions. The high sensitivity of currently available balances permits study of very low reaction rates at which mass transfer limitations are minimized. Both low- and high-pressure thermobalances have been described in the literature [11] and have been used recently in coal gasification experiments [17,18].

The data obtained in gravimetric studies cannot be related to other experimental work unless the samples tested are carefully characterized in terms of surface area and pore size before and during testing. This is clearly shown by the work of Otto and Shelef using an atmospheric pressure balance [6,17]. In a comparison of a number of different coals and chars with graphite, they found considerable differences in the gasification rates per unit weight of sample, some samples gasifying up to two thousand times as fast as did graphite. On the other hand, the rates based on the surface area of the samples were found to be within one order of magnitude.

As the gasification progressed and the char was consumed, the surface area increased by a factor of four. In spite of this change in the area, the gasification rate per unit area remained constant.

Characterization of the coal is particularly important in catalytic studies, since it allows discrimination between physical and chemical effects, both of which can occur in gasification [19]. The determination of surface area of carbons and coal is not simple. The surface area varies considerably with the level of pyrolysis and devolatilization. Coal has extensive microporosity which can account for up to 80% of the pore volume of a high-ranking coal [20]. Some of the pores are in the 5–8-Å range leading to severe diffusion limitations and inaccuracies in the surface area determination if the adsorption measurements are carried out at −196°C [17]. The micropore structure opens up as the gasification consumes carbon.

18.2.2 Catalysts

The contacting problem in gasification has been surmounted in processes currently under development by the use of molten salts, among which the most common are K_2CO_3 and Na_2CO_3. The low melting points (891 and 851°C, respectively) and relatively low cost of these salts make them excellent candidates for disposable gasification catalysts. Their effect on the combustibility of carbon has been known for more than fifty years [21,22]. Basic materials are used extensively by the petrochemical industry to enhance the reactivity of carbon with steam. For example, in steam reforming of hydrocarbons, alkali and alkaline earth oxides are used as catalyst promoters to minimize coking [23], potash being considered one of the best promoters for this purpose.

Battelle tested the effect of impregnating coal with salts [19], and found enhanced gasification with sodium and potassium salts. The observed effect may have been partly physical due to an increase in the reactivity of the coal caused by the impregnation procedure itself.

Interestingly, alkaline materials may be very convenient catalysts for coal gasification because of their presence in certain coals and their abundance in a number of minerals.

Shelef and Otto have compared the gasification rates of various coals with that of graphite [6,17]. While all noncatalyzed rates were found to be within one order of magnitude, Braunkohle lignite stood out as having the highest rate. The mineral matter extracted from this coal catalytically enhanced the gasification rates of less reactive coals [24]. A comparison of the composition of the mineral matter in Braunkohle with that of other coals shows low levels of Si, Al, K, and Ti, and a very high level of Ca. Whether a Ca compound is indeed the gasification catalyst is not definitely established. It seems that Ca is not an effective additive to prevent C deposition in steam reforming, in particular when compared to K [24]. Calcium in combination with sodium has been used successfully by Battelle in their impregnation process [19], and has been claimed in several patents to be a gasification catalyst [25].

There are many minerals that contain Na and K and could yield catalysts for coal gasification with minimal processing. Their utilization should be considered. An example for such materials is the brine from Searles Lake [26] which is being used for the production of potassium and contains considerable amounts of Na_2CO_3 and other Na salts which may not have to be removed for gasification applications.

Steam is an effective vapor transport agent for alkali salts [27] and could help to redisperse alkaline catalysts continuously as the coal is consumed (see also Section 18.2.5).

18.2.3 Preparative Procedures

Impregnation of coal with a catalyst has, in principle, the same objective as the preparation of supported catalysts, where maximum dispersion of the active species on the high surface area support is desired. Many factors affect the preparation of such high surface area materials (Chapter 5). They include the pH of the impregnating solution, the nature of the salt, the concentration of the solution, and the mode of impregnation. There are many different impregnation techniques, some of which may be especially suited for impregnation of coal. Because of the extensive microporosity of the coal, the use of vacuum impregnation may be very effective and may result in much better penetration. However, the improvement in the gasification would have to be substantial to justify such a complicated coal treatment procedure. Changes in pH of the impregnating solution may also be very important in view of the acid–base nature of coal [28].

The pretreatment of the coal is expected to be another important parameter in catalyst impregnation. It is likely to influence not only the possible interaction of the catalyst with the coal but the rate of gasification as well, by changing the pore structure and internal surface area. As the gasification proceeds, the microstructure of the coal opens up and the surface area increases.

In view of the effect of pretreatment and surface area on the gasification rate, means should be considered to maximize the effect. In accomplishing this, rapid devolatilization, the degree and rate of pyrolysis, and, in general, the history of the coal before gasification may be important. Increase in surface area may be obtained by rapid heating of highly volatile liquids preadsorbed on the coal. A rapid vaporization at a rate exceeding the diffusion rate out of the microstructure will result in breakup of the coal matrix. In the most dramatic example of this phenomenon, molecules intercalated within the planes of layered compounds open the structure to such a degree that there is a macroscopic change in the material. The effect, known as exfoliation, is used commercially for clays such as vermiculite to achieve a very open structure. It has also been observed for graphite, which has a layered structure. The possibility of taking advantage of this effect in the case of coal should be explored, in view of the layered arrangement of coal structure models. Since the gasification rate is proportional to the surface area, an increase in area by exfoliation would have a significant effect on the gasification process.

18.2.4 Mechanism of Gasification

While there is only limited information on the mechanism of the catalytic reaction between steam and coal, the effect of metals on the reaction of oxygen with carbon, and in particular with graphite [15,29,30], is better understood. In this system the rate is critically dependent on the structure of the sample, the edges of the hexagonal planes of graphite being the most reactive [16]. The role of the metal is to dissociate oxygen molecules and to provide reactive oxygen atoms. As gasification proceeds, the carbon in contact with the catalyst is consumed, and certain catalyst particles become mobile, their mobility depending on the environment [31].

The catalyst mobility and the continuously changing catalyst–coal interface in all gasification reactions present problems in the interpretation of gasification data. The dynamic nature of the system has to be incorporated in any model of the gasification reaction, and should also be included in the consideration of improved catalyst systems. The degree of mobility and the balance between redispersion and agglomeration of catalytic particles at the conditions of gasification may determine catalytic activity.

Because of the mobility of the catalyst and the effect of surface area and pore size on gasification activity, measurement of the physical properties of the coal at different stages of gasification may be an effective means to monitor the reactivity of various materials.

The basicity of the catalyst surface is likely to be important in determining the interaction of water with the catalyst and the coal and in influencing the reactivity of coal itself. The mechanism of catalysis of the steam gasification by metals may be similar to that established for oxidation of graphite inasmuch as the metal oxide is the source of dissociated oxygen. However, the action of other materials such as K_2CO_3 may be very different and both systems should be examined.

The study of the mechanism of catalytic coal gasification is an important supporting program. It should include such aspects as the effects of pretreatment and of catalysts on the properties and reactivity of coal and on its devolatilization.

18.2.5 Exploratory Programs

In this section two ideas for the development of new catalytic gasification methods will be discussed. One concerns the intercalation of layered structures; the other, volatile catalysts.

The similarity between the structures of graphite and coal suggests that the interesting phenomenon of intercalation observed on graphite may be applicable to coal gasification.

To understand the nature of intercalation, it is useful to refer to the structure of graphite. A graphite crystal is composed of layers of hexagonal carbon networks. The carbon–carbon distance within these networks is 1.42 Å. The distance of 3.35 Å between layers indicates very weak interlayer forces which permit one layer to slide over the other, and are responsible for the unique lubricating properties of graphite. Because of this structure, graphite is able to accommodate various compounds between the layers which interact strongly with the guest molecules. Intercalation occurs also with a number of other layered compounds such as micas and, in particular, transition metal sulfides, selenides, and tellurides.

The examples of graphite intercalation compounds shown in Table 18-2 include compounds with electron donors (K) and with electron acceptors (Br). The strength of the interaction varies with the nature of the intercalating species and the stoichiometry. It can be substantial, as in the case of the potassium–graphite system where the heat of intercalation ranges from 7 to 18 kcal mol^{-1}. This strong interaction results in changes in the properties of the graphite. Thus, upon potassium intercalation, the conductivity changes by a factor of 10 [32], and the interlayer spacing expands to 5.40 Å [33]. The expansion can be as high as 9.45 Å in the case of C_nFeCl_3, and, in other layered compounds such as TaS_2, expansions of interlayer spacings to 50 Å have been observed [34].

Because of the changes in interlayer spacing, the adsorption properties of the graphite change as well. The compound $C_{24}K$, for example, adsorbs large volumes of hydrogen until the stoichiometry $C_{24}KH$ is reached [35]. This compound and a number of other graphite intercalates exhibit catalytic activity in a variety of reactions [33], including low-temperature ammonia synthesis [36] and hydrogenation of unsaturated hydrocarbons [37]. Since the properties of graphite are changed by intercalation it is likely that its reactivity changes as well. Intercalation may, therefore, be an effective way to overcome the catalyst–carbon contacting problem in reactions such as gasification. The challenge is to find an interca-

TABLE 18-2
SELECTED INTERCALATION COMPOUNDS OF GRAPHITE

C_6Li, $C_{12}Li$, $C_{18}Li$	C_nFeCl_3
$C_{64}Na$	$C_m^+[GaCl_4^- + nGaCl_3]$
C_8K, $C_{10}K$, $C_{24}K$, $C_{36}K$	$C_m^+[AlCl_4^- + nAlCl_3]$
C_8Rb, $C_{24}Rb$	$C_{12}SbCl_5$
$C_{12}M(NH_3)_2$ (M = Li, Na, K, Rb, or Cs)	$C_{16}N_2O_5$
$C_{12}Li(CH_3NH_2)_2$	C_7CrO_3
C_8Br, $C_{16}Br$	

lated system that will remain stable at high gasification temperatures and inert to the sulfur and ash compounds of coal.

A second area that is of interest for long-term research is the use of catalysts that are volatile at the conditions of gasification. The use of steam as a potential vapor transport and redispersing agent for alkaline catalysts has been mentioned in Section 18.2.2. Recent work with coal chars [24] shows that Ru, which possesses a volatile oxide, retains its catalytic activity for the gasification of coal much longer than other transition metals. This same volatility could be used to remove the catalyst from the ash and retain it in the reacting coal bed, were it not for the high cost of Ru and its susceptibility to sulfur poisoning which of course make it an impractical catalyst. A search for appropriate volatile catalysts or volatilization methods seems worthwhile.

18.3 CONCLUSIONS

The gasification of coal faces a number of restricting constraints in the choice of catalysts for this system. They include such catalyst problems as contacting with coal, deactivation by sulfur or mineral matter, recovery, and regeneration. Because of these severe constraints, the most likely catalysts for coal gasification are inexpensive, disposable materials that will be used on a once-through basis.

In view of the limited amount of information available concerning the role of the catalyst and the nature of its activity, the major effort in this area is short range. Effective testing programs using thermogravimetric techniques can lead to a better understanding of the mechanism of the catalytic reaction and the determination of catalyst properties important for activity. Characterization of the coal before and during reaction is crucial for this purpose since surface area and pore size change as the coal is being consumed. The gasification rate is a function of coal surface area, and measurements of the area of the sample have to be included in any data analysis.

The pretreatment of the coal can affect the surface area, the porosity, and the contacting of the catalyst with the coal during impregnation. These aspects need careful study.

Impregnation of coal appears to be an effective method for incorporating a gasification catalyst. In view of the obvious similarity between this procedure and the preparation of supported catalysts, many of the catalytic techniques may be applicable to coal gasification. Research in this area appears justified.

REFERENCES

1. Lowry, H. H., ed., "Chemistry of Coal Utilization," p. 895. Wiley, New York, 1963.
2. Thomas, J. M., and Walker, P. L., *Carbon* **2**, 434 (1965).

3. Tomita, A., Sato, N., and Tamai, Y., *Carbon* **12**, 143 (1974).
4. Willson, W. G., Sealock, L. J., Jr., Hoodmaker, F. C., Hoffman, R. W., Stinson, D. L., and Cox, J. L., *in* "Coal Gasification" (L. G. Massey, ed.), p. 203. American Chemical Society, Washington, D.C., 1974.
5. Haynes, W. P., Gasior, S. J., and Forney, A. J., *in* "Coal Gasification" (L. G. Massey, ed.), p. 179. American Chemical Society, Washington, D.C., 1974.
6. Otto, K., and Shelef, M., *Proc. Int. Congr. Catal., 6th* (G. C. Bond *et al.*, eds.), p. 1082. The Chemical Society of London, 1977.
7. Otto, K., private communication, 1975.
8. Boudart, M., Cusumano, J. A., and Levy, R. B., New Catalytic Materials for the Liquefaction of Coal, Rep. RP-415-1, Electric Power Research Institute, October 30, 1975.
9. Cox, J. L., Sealock, L. J., and Hoodmaker, F. C., *Energy Sources* **2** 83 (1975).
10. Kertamus, N. J., and Woolbert, G. D., *Energy Sources* **2**, 203 (1975).
11. Squires, A. M., *in* "Fuel Gasification" (R. F. Gould, ed.), p. 205. American Chemical Society, Washington, D.C., 1967.
12. Snider, R. B., Wilson, W. I., Vogel, G. J., and Jonke, A. A., Sulfation and Regeneration of Synthetic Additives, presented at the *Int. Conf. Fluidized-Bed Combust., 4th, McLean, Virginia* December 9–11, 1975.
13. Bienstock, D., and Field, F. J., *J. Air Pollut. Contr. Assoc.* **10**, 121 (1969).
14. Ode, W. H., *in* "Chemistry of Coal Utilization" (H. H. Lowry, ed.), p. 209. Wiley, New York, 1963.
15. Toropov, N. A., Handbook of Phase Diagrams of Silicate Systems, Vol. I: Binary Systems, translated from Russian NTIL-TT71-50040, 1972.
16. L'Home, G. A., and Boudart, M., *Symp. Combust., 11th* p. 197, 1967.
17. Otto, K., and Shelef, M., *Am. Chem. Soc., Div. Ind. Eng. Chem., Symp. Catal. Convers. Coal, Pittsburg, Pennsylvania* April 1975.
18. Gardner, N., Samuels, E., and Wilks, K., *in* "Coal Gasification" (L. G. Massey, ed.), p. 217. American Chemical Society, Washington, D.C., 1974.
19. Chauhan, S. P., Feldmann, H. F., Stambaugh, E. P., and Oxley, J. H., *Am. Chem. Soc. Div. Fuel Chem. Prepr.* **20**, 207 (1975).
20. Van Krevelen, D. W., "Coal," p. 143ff. Elsevier, Amsterdam, 1961.
21. Taylor, H. S., and Neville, H. A., *J. Am. Chem. Soc.* **43**, 2055 (1921).
22. Blayden, H. E., Riley, H. L., and Shaw, F., *Fuel* **22**, 64 (1943).
23. Rostrup-Nielsen, J. R., "Steam Reforming Catalysts." Teknisk Forlag A/S, Copenhagen, 1975.
24. Shelef, M., private communication, 1975.
25. Squires, A. M., *in* "Fuel Gasification," p. 104. American Chemical Society, Washington, D.C., 1967.
26. Kent, J. A., ed., "Riegel's Handbook of Industrial Chemistry," 7th ed., p. 548. Van Nostrand–Reinhold Co., Princeton, New Jersey, 1974.
27. Schafer, H., "Chemical Transport Reactions," p. 50. Academic Press, New York, 1974.
28. Sternberg, H. W., Raymond, R., and Schweigherdt, F. K., *Science* **188**, 49 (1975).
29. Thomas, J. M., *in* "Chemistry and Physics of Carbon" (P. L. Walker, Jr., ed.), Vol. 1, p. 121–202. Dekker, New York, 1966.
30. Hennig, G. R., *in* "Chemistry and Physics of Carbon" (P. L. Walker, Jr., ed.), Vol. 2, p. 1–43. Dekker, New York, 1966.
31. Baker, R. T. K., France, J. A., Rouse, L., and Waite, R. J., *J. Catal.* **41**, 22 (1976).
32. Blackman, L. C. F., Matthews, J. F., and Ubbelohde, A. R., *Proc. Roy. Soc. London* **A258**, 339 (1960).
33. Boersma, M. A. M., *Catal. Rev.-Sci. Eng.* **10**, 243 (1974).

34. Geballe, T., *Sci. Am.* **225** (5), 22 (1971).
35. Watawabe, K., Kondow, T., Soma, M., Onishi, T., and Tamaru, K., *Proc. Roy. Soc. London* **A333**, 51 (1973).
36. Ichikawa, M., Kondo, T., Kawase, K., Sudo, M., Onishi, T., and Tamaru, K., *Chem. Commun.* 176 (1972).
37. Ichikawa, M., Soma, M., Onishi, T., and Tamaru, K., *J. Catal.* **9**, 418 (1968).

Chapter 19

Synthesis of Fuels and of Selected Feedstocks from Carbon Monoxide and Hydrogen

19.1 INTRODUCTION

This chapter covers primarily the synthesis of hydrocarbon fuels from CO–H_2 mixtures by the Fischer–Tropsch and related reactions, but touches also upon the synthesis of oxygenated compounds. In some variations of the Fischer–Tropsch reaction, oxygenated compounds form in smaller or larger quantities, even if undesired. They can be used directly as fuel or as feedstocks for hydrocarbon synthesis.

Depending on the intended use of the Fischer–Tropsch products, their desired type and distribution can very widely. For use as chemical feedstocks, ethylene, propylene, benzene, toluene, and xylenes are highly desirable. In the production of fuels for facilities combining coal gasification and electric power generation, selectivity is not essential, and practically any liquid hydrocarbon is acceptable. For the production of motor fuels, isoparaffins in the range of C_6 to C_{10} and single-ring aromatics are preferred. It is apparent that there is a need for selectivity control in producing the extensive range of desired hydrocarbon products. In actual operation, quantities of methane and of other less desirable materials are produced or product distributions are obtained that are very broad, with a significant fraction of the product below or above the required range. with a significant fraction of the product below or above the required range.

The thermodynamics of hydrocarbon synthesis from CO and H_2 has been described extensively in the literature [1,2]. Review of these data lead to these observations important for Fischer–Tropsch reactions:

(a) All of the reactions forming hydrocarbons from H_2 and CO are exothermic; the free energy changes of the reactions increase and the equilibrium constants decrease with temperature.

(b) The formation of methane has the most negative free energy change per carbon atom.

(c) Below 350°C most of the reactions have equilibrium constants above unity, while above 450°C nearly all of the synthesis reactions become thermodynamically unfavorable. The exceptions to these generalizations are the formation of methane, for which the equilibrium constant exceeds one up to 650°C, and the formation of acetylene and methylacetylene, which have equilibrium constants less than one at temperatures as low as 200°C.

(d) The hydrogenation of CO_2 to hydrocarbons is thermodynamically more favorable than the hydrogenation of CO.

(e) Reactions of CO and H_2 with methane to form higher hydrocarbons are thermodynamically feasible.

(f) The formation of olefins is slightly less favored than that of paraffins. Reactions to cyclohexane and methylcyclohexane have equilibrium constants very near those for *n*-hexane and *n*-heptane.

(g) The formation of isomeric paraffins is thermodynamically favorable in comparison to that of normal paraffins.

(h) The free energies of formation of aromatic hydrocarbons are very close to those of the monoolefins. The formation of aromatics from CO and H_2 is also possible below 400°C.

(i) Ethanol and higher alcohols can be produced in substantial quantities at synthesis conditions.

(j) Hydrogenation of olefins and dehydration of alcohols are thermodynamically possible at usual synthesis conditions. Thus, alcohols must be a primary product while olefins and paraffins may be produced by dehydration of alcohols and hydrogenation of olefins, respectively.

(k) An increase in operating pressure will increase the conversion of the Fischer–Tropsch reactions as a result of the large volume decrease associated with the synthesis reactions.

The implications of the thermodynamic observations are that low reaction temperature and increased pressure is preferred in the synthesis of higher hydrocarbons. Since methane is by far the preferred equilibrium product, kinetic control and selectivity must regulate the product distribution. Generally, the product distribution observed during synthesis differs substantially from that which corresponds to thermodynamic equilibrium. For example, straight-chain paraffins and α olefins represent a large fraction of the hydrocarbons produced. Only some isoparaffins and virtually no cyclic or aromatic materials are produced. Carbon monoxide is preferentially hydrogenated due to its stronger interaction with the surface, and CO_2 reacts only after most of the CO has been removed.

The general Fischer–Tropsch reactions for paraffin formation can be written as

$$(2n + 1)H_2 + nCO \rightarrow C_nH_{2n+2} + nH_2O \quad (19\text{-}1)$$

$$(n + 1)H_2 + 2nCO \rightarrow C_nH_{2n+2} + nCO_2 \quad (19\text{-}2)$$

depending on whether water or carbon dioxide is the byproduct. Similar reactions can be written for olefins, aromatics, cycloalkanes, and alcohol production. In general, at the low temperatures usually employed in the synthesis reaction, water is the primary product and CO_2 a secondary product formed as a result of the water–gas shift reaction, which is slow over most Fischer–Tropsch catalysts. The ratio of H_2 to CO required in the feed to the synthesis catalyst is approximately 2:1. Varying the H_2/CO ratio may have an important effect on the product distribution.

19.2 MAJOR PROBLEMS AND CATALYST DEVELOPMENT OBJECTIVES

Hydrocarbon synthesis catalysts suffer from the same problems as methanation catalysts: sulfur poisoning, carbon deposition, and thermal degradation. Control of product selectivity is desirable as is high activity since low-temperature operation appears to be necessary for optimum selectivity.

Much of the older work on Fischer–Tropsch synthesis was done with iron and cobalt catalysts containing a variety of promoters and covered the effects of temperature, pressure, residence time (space velocity), H_2/CO ratio, and sulfur level. Product selectivity was reported mainly in terms of distillation fractions. In some cases, water-soluble organics were also measured separately. The lack of reliable data on specific products was due primarily to the unavailability of modern chromatographic and mass spectrometer instrumentation necessary for complete and rapid analysis. More recent data generally report results for specific products from methane to higher paraffins and include isoparaffins, alkenes, and alcohols. Detailed product analysis is important not only in the testing of fresh catalysts but also in assessing degradation since selectivity may change substantially.

19.2.1 Selectivity

An extensive review of the product composition from a variety of catalysts is given by Anderson [1, p. 109] who discusses the effects of promoters and process variables.

An early catalyst system that was optimized to give good activity and selectivity included supported cobalt catalysts having the composition of 100Co:18ThO_2:100kieselguhr and 100Co:5ThO_2:10MgO:200kieselguhr. A typical product distribution obtained at the conditions given on the former catalyst is presented in Table 19-1.

The factors that influence the product distribution for such Co catalysts are these:

(a) At 160–175°C, and low conversions of CO and H_2, a significant fraction of the product consists of oxygenated compounds, predominantly alcohols. At 190°C the alcohol fraction is very small.

(b) The olefin content varies from ~25% at C_5 to 10% at C_{10}, and continues to decrease as the chain length increases.

(c) Operation at 5–15 atm shifts the product distribution to higher molecular weights, decreases the olefin content, and increases the fraction of wax (> C_{18}).

(d) Promoters such as thoria and potassium carbonate have only a small effect on the selectivity. Thoria, however, increases activity of Co catalysts

TABLE 19-1

FISCHER–TROPSCH SYNTHESIS WITH VARIOUS CATALYSTS

	Catalyst no.		
	I[a]	II[b]	III[c]
Process conditions			
Temperature (°C)	190	190–210	315
Space velocity (per hour)	100	250	700
H_2/CO ratio	2:1	1.3:1	2:1
Pressure (atm)	1	20	18
Reactor	Packed	Packed	Fluidized
Recycle ratio	—	10	2
Product composition (%)			
C_1	13	12	9
C_2	2	6	8
C_3–C_4	9	25	30
C_5–C_8	25	22	32
C_9–C_{16}	35	17	15
C_{17}–C_{19}	6	18	6
C_{20+}	10		
Reference	[3]	[1, p. 213ff]	[4]

[a] 100 Co:18ThO_2:100 kieselguhr.
[b] Fe_3O_4–Al_2O_3–K_2O (fused).
[c] Fe-oxide–K_2CO_3 (reduced).

permitting the use of lower temperatures with a greatly preferred product distribution.

(e) On increasing the temperature, the average molecular weight of the products decreases and the fraction of gaseous products, and especially of methane, increases.

(f) Decreasing the H_2/CO ratio from 2 to 1 greatly increases the olefin content, the average molecular weight, and the alcohol production. The ratio of H_2 to CO consumed remains near 2.

The second example is a fused iron catalyst operated in a fixed-bed reactor with gas recycle (Synol Process) [1, pp.213ff], a process optimized to produce oxygenated products. Its characteristics, operating conditions, and product distribution are given in Table 19-1.

The following observations were made on this catalyst. Operation at atmospheric pressure generally results in poor activity, short life, and a product distribution skewed to low carbon numbers. Alcohols form up to high carbon number and represent a relatively large fraction of the total product. The olefin fraction is also very large. Compared with the product distribution on Co, that on iron has a sharper maximum which appears at lower carbon number. Iron also gives more branched hydrocarbons. The relatively large fraction of high molecular weight materials, C_{17+}, is not a general property of Fe catalysts.

A second Fe catalyst, for which there is a large body of very detailed data, is a K_2CO_3-promoted iron oxide [4]. A typical product distribution for this catalyst is presented in Table 19-1 at the operating conditions given. With this catalyst, 27% of the total products were oxygenated compounds, and there was also a relatively large portion of olefins, especially at low carbon numbers. The olefin content was shown to be a function of the alkali content, and catalysts containing no alkali produced very little olefinic material.

The reviews of Anderson [1] and of Storch *et al.* [2] cover much of the earlier work, and Pichler and Schulz [5] and Eidus [6] most of the intervening work. The number of possible additives and reaction conditions that affect operation of the catalyst make the system extremely complicated and at present not well understood.

Various reactor types have been tried, including recirculating oil [1], hot gas recycle [7], and the fluidized-bed reactors mentioned earlier. These investigations were directed at achieving close temperature control to maintain product selectivity.

The only commercial Fischer–Tropsch plant in operation is in Sasolburg, South Africa. It is optimized for the production of liquid hydrocarbons. However, the fluidized-bed Synthol reactors currently in use produce 14% methane, and their $C_1 + C_2$ fraction represents 27% of the total hydrocarbon production [5,6]. The product distribution is considered quite good with the useful hydrocarbon fraction, defined as C_{3+}, representing 64% of the products. The heavy hydrocarbon wax portion is very small due to the operating temperature and the recirculating catalyst. Further tightening of the product distribution will require a greater understanding of the reacting system.

Basically, the mechanism is the same as that for methane synthesis with the additional step of polymerization of the hydrocarbon intermediates on the surface to liquid hydrocarbons. The hydrocarbon chain length is determined by a competition between the polymerization of the surface species and the hydrogenation and desorption of the surface species and the hydrogenation and desorption of the surface polymer in a termination step. An alternate mechanism involves the repeated insertion of CO into a surface hydrocarbon residue. Either mechanism is consistent with many of the observed effects of catalyst composition and process variables on product selectivity.

Alteration of the catalyst to affect the relative binding of CO and H_2 will change the surface coverage of the surface complex and the probability of subsequent carbon addition. Indeed, promoters such as K_2O which are strong bases have been shown to shift the hydrocarbon distribution to longer chain length. Potassium oxide is thought to donate electrons to the metal thereby strengthening of the CO—metal bond, and also weakening the C═O bond. Metals with lower hydrogenation activity generally give higher molecular weight product distributions and larger proportions of olefins. A change in H_2/CO ratio affects the surface composition of adsorbed H_2 and CO and the resulting product distribution.

19.2.2 Sulfur Poisoning

Conventional synthesis catalysts are highly sensitive to sulfur poisoning at very low levels of sulfur. The interaction of H_2S with the catalyst is similar to that encountered in methanation with nearly quantitative uptake of the sulfur and complete deactivation of the catalyst. Karn and co-workers [8] studied the sulfur poisoning of several iron catalysts in a bench scale test at high pressure and high conversion. Sulfur levels, in the range of 5–50 ppm H_2S, led to very rapid deactivation of the catalysts. The activity decreased approximately linearly with sulfur uptake and approached zero when the sulfur level on the catalyst reached one monolayer. Most of the sulfur fed to the reactor was retained by the catalyst even at levels of 50 ppm H_2S, and at 350°C [9]. At 15 ppm H_2S, a massive iron catalyst, the surface area of which is predominantly reduced iron, can be completely poisoned in 100 h of operation. A useful catalyst life with such catalysts would require very low H_2S levels, around 0.1–0.01 ppm (see Section 17.2).

Product selectivity also changes as the catalyst is poisoned by sulfur. Increased methane production is observed on promoted iron extensively deactivated by sulfur addition [8]. Herington and Woodward [10] found that sulfur addition to a thoria-promoted cobalt catalyst first increased the liquid hydrocarbon yield, then more extensive poisoning shifted the product distribution to gaseous hydrocarbons. Similar results are reported for Ni and promoted Ni catalysts [1, p.246]. Craxford attributed the initial shift in product distribution toward liquid hydrocarbons to the poisoning of the hydrocracking function of the catalyst [11]. Herington and Woodward ascribed the effect to a preferential poisoning of the surface hydrogenation function, thus leading to longer hydrocarbons due to increased polymerization [10].

An additional problem of sulfur poisoning is that a partially deactivated catalyst cannot be operated at higher temperature to maintain conversion. Such temperature increases will adversely affect the product distribution and, if excessive, will cause carbon deposition and plugging of the catalyst bed.

19.2.3 Carbon Deposition

The deposition of carbon by the Boudouard reaction

$$2CO \rightleftharpoons CO_2 + C \tag{19-3}$$

can lead to catalyst deactivation, destruction of the catalyst pellet, and increased attrition and physical plugging of the reactor. The conditions under which carbon formation is thermodynamically permitted are well defined (Section 17.2.3). The ternary diagram of White *et al*. [12] is a particularly useful presentation format and permits a rapid assessment of whether carbon deposition is thermodynamically favored. However, these thermodynamic calculations assume the establishment of the equilibria in the reactions involving H_2 and H_2O:

$$H_2O + CO \rightleftharpoons H_2 + CO_2 \tag{19-4}$$

$$3H_2 + CO \rightleftharpoons H_2O + CH_4 \tag{19-5}$$

$$H_2O + C \rightleftharpoons H_2 + CO \tag{19-6}$$

At conditions where kinetic limitations prevent these reactions from proceeding to equilibrium, the above-described thermodynamic predictions became tenuous.

At the usual conditions employed in Fischer–Tropsch synthesis, i.e., an H_2/CO ratio less than 2, pressures of 1–50 atm, and temperatures below 350°C, carbon is thermodynamically stable [13]. However, it is generally found that very little carbon is deposited on most Fe, Ni, and Co catalysts at temperatures below 300°C. As the temperature is raised above this limit, carbon deposition has been observed to occur on iron lathe turnings [14] and on sulfur-poisoned fused-iron catalysts [8]. The absence of carbon deposition at low temperatures is a result of the low rate of the Boudouard reaction on these catalysts.

19.2.4 Activity Maintenance

Other processes which cause activity loss include physical and chemical changes in the catalyst, adsorption of high molecular weight products, and attrition. In the synthesis of hydrocarbons, even slightly degraded activity can have an adverse effect on selectivity directly, or indirectly by requiring higher operating temperature in order to maintain the reaction rate.

The latter effect is exemplified by the commercial operation at Sasolburg, South Africa, where promoted iron is used as a catalyst and generally replaced after approximately 50 days [15]. The synthesis gas is freed of sulfur down to 0.01 ppm, and should not cause significant problems. However, in use, the catalyst deteriorates and the temperature is raised to maintain a constant synthesis rate. The higher operating temperature results in a change in product distribution toward lower carbon numbers, and catalyst life becomes a trade-off between catalyst replacement cost and the poor product selectivity. Pichler considers the same phenomenon with Co catalysts that show a slow deterioration with time [5]. This is partly due to sulfide and sulfate formation and partly to carbon deposition and sintering caused by inadequate control of temperature.

19.2.5 Synthesis of Diesel Fuel, Jet Fuel, and Liquefied Petroleum Gas (LPG)

The Fischer–Tropsch products are well suited for diesel fuel, jet fuel, and LPG. These particular materials have recently received attention due to their national importance [16]. The properties and chemical composition of diesel and jet fuels were reviewed earlier (Chapter 12).

Liquefied petroleum gas consists essentially of C_3 and C_4 hydrocarbons. It is, in fact, sold in many cases as butane or propane with minimum concentrations of the major component specified. This composition gives LPG a very high energy content compared to methane, 2500 Btu ft^{-3} for propane and 3200 Btu ft^{-3} for butane, making this material very useful as an exceptionally clean fuel in mobile equipment (octane number of 100) as well as a storable replacement for natural gas. The sulfur content depends on use, but is generally low.

In this class of fuels the preferred components are paraffinic hydrocarbons, mainly normal paraffins with some branched chain components desirable to control physical properties. These are just the materials produced by the low-temperature Fischer–Tropsch synthesis as practiced over promoted iron and cobalt catalysts. The product distribution for a typical cobalt catalyst (Table 19-1) shows a broad maximum from C_5 to C_{16}. This is somewhat low for diesel fuel where the distillation range of 325–800°F specifies a hydrocarbon composition extending from approximately C_9 to C_{25}. Thus, much of the synthesis products is below the desired range. A lighter distillation fraction such as kerosene for use as a jet fuel has a distillation range of 300–500°F and corresponding composition in the vicinity of C_9 to C_{16}. This more closely corresponds to that found on the Co catalyst. The important point is that appropriate fractions of the synthesis products formed under certain operating conditions can be used for this class of fuels with minimum subsequent treatment. The broad range of paraffinic hydrocarbons together with a small fraction of α olefins and branched chain hydrocarbon may only require a simple hydrogenation step and distillation. If the heavier diesel fuel fraction is wanted, a shift in the product distribution to higher molecular weights would be required. Alternately, the product distribution obtained from the fused promoted iron catalysts (Table 19-1) has a maximum of C_3 and C_4 material, necessary for LPG. However, a majority of the total product is outside of this range.

19.3 RESEARCH AND DEVELOPMENT AREAS

Application of recent developments in catalysis and related disciplines to the solution of problems encountered in the synthesis of hydrocarbon fuels and chemicals from CO and H_2 will be considered here under three headings: short-term, long-term, and basic research and development programs.

19.3.1 Short-Term Programs

19.3.1.1 Examination of Common Catalysts

In the search for good yields of liquid hydrocarbons or specific chemicals, a variety of catalysts must be evaluated for activity and selectivity. The major portion of past work was concerned with Co, Fe, Ni, and to a smaller extent with Ru and combinations of these metals with promoters. Pt, Pd, and other noble metals have not been examined in detail under varied conditions for activity and selectivity. One example which shows the possible advantages of these materials is the use of alumina-supported Pt to produce large fractions of ethane and dimethyl ether reported recently by Vannice [17]. At atmospheric pressure and with a H_2/CO ratio of 2, Pt shows very high selectivity to methane [18]. At lower H_2/CO ratios, and especially at higher pressures, the production of ethane increases. As the pressure is raised to 10 atm or higher, the fraction of dimethyl ether increases dramatically, as is shown in Table 19-2.

The data presented above emphasize the importance of the reaction conditions, especially H_2/CO ratio and total pressure in controlling product distribution. Although most common catalytic materials have been examined for CO–H_2 synthesis in one form or another, the systematic investigation of a variety of these materials may provide useful leads on catalyst properties necessary for the pro-

TABLE 19-2

SYNTHESIS OF ETHANE, ETHYLENE, AND DIMETHYL ETHER ON Pt/Al_2O_3[a]

Temperature (°C)	Pressure (atm)	H_2 ratio	Product distribution (mol%) C_1	C_2	C_3	$(CH_3)_2O$
280	1.0	2	90+	5	0	0
271	1.0	0.5	82	19	0	0
263	1.96	1	89	12	0	0
271	2.12	1	85	10	0	5
274	10.1	1	70	8	0	23
269	30.0	1	67	4	0	30

[a] From Vannice [17].

duction of specific chemicals. Changes in activity and selectivity with time should also be investigated since changes in the bulk or surface composition of the catalytic phase during synthesis is the rule rather than the exception.

The support is an important aspect of the supported metal catalyst and should also be examined. For example, the design of a bifunctional catalyst system with a hydrogenation component and an isomerization component may be a promising route to various chemicals and valuable fuels. The importance of the support in stabilizing highly dispersed supported metals, and in modifying the properties of the metal through metal–support interaction, has been repeatedly emphasized and should be given consideration in research efforts.

Other catalytic materials of interest for controlling the activity, selectivity, and activity maintenance properties of catalysts for Fischer–Tropsch reactions include bimetallic alloys or clusters. The properties of these materials are now better understood and modern spectroscopic and adsorption techniques permit one to fully characterize the surface and bulk chemical and physical properties of these systems. These catalysts are of primary interest because, by addition of one metal to another, one can, in principle, control various catalytic properties over very extensive ranges. One group of bimetallic catalysts which have been widely studied and are of particular interest for synthesis reactions involves a combination of Group VIII (e.g., Rh, Ru, Ir, and Ni) and Group IB (Cu, Ag, or Au) metals. These combinations enable one to control hydrogenolysis and hydrogenation activity [19] which are thought to be of importance in determining selectivity in the Fischer–Tropsch reaction. The behavior and properties of such bimetallic catalysts are fairly well understood, and their application to hydrocarbon synthesis appears promising.

19.3.1.2 Poison-Resistant Catalysts

The need to remove sulfur to the low levels required by many hydrocarbon synthesis catalysts (0.1 ppm) increases the cost of the products and reduces the efficiency at which the process can operate. The existence of a catalyst capable of operating in a feedstream with sulfur levels in the range of 10–100 ppm would be advantageous since the end use of the hydrocarbon products can, in most cases, tolerate such levels of sulfur. The problem in the present case is much more complicated and difficult than that encountered with methanation where use can be made of the high thermodynamic stability of methane and of the high operating temperatures to control selectivity. Hydrocarbon synthesis catalysts are optimized to give a desired product distribution, and introduction of sulfur into the system will adversely alter the selectivity.

Controlled sulfur poisoning can be beneficial to selectivity by shifting the product distribution toward higher hydrocarbons and decreasing the production of methane [1, p.246; 7; 10; 20]. The mechanism of operation is thought to be a poisoning of the surface hydrogenation and hydrogenolysis activity leading to an increased hydrocarbon chain length. Of course, if the effect of sulfur could be moderated, serious loss in total activity could be avoided. So far few catalysts have been examined for product distribution in a sulfur-containing stream, and an effort should be made to extend such studies.

The materials normally considered as catalysts for hydrocarbon production, Fe, Co, Ni, and the various promoted mixtures of these should be tested initially, if only to establish a baseline. A systematic study of the important operating parameters on several representative catalysts, including variation in the level of H_2S, should provide an adequate background. Detailed product analysis would be necessary to permit an assessment of changes in product selectivity.

Materials normally not considered good catalysts should then be examined. The deactivating and selectivity moderating properties of sulfur could have beneficial effects on the catalytic characteristics of some materials (Section 6.2). Platinum, for example, is thought to bind CO too strongly, leading to low activity [21], and to owe its high selectivity to its good hydrogenation activity. Sulfur poisoning will certainly affect both of these properties, but the resulting activity and product distribution can be obtained only by experiment. Platinum is also much less susceptible to bulk sulfidation than are Fe, Co, or Ni [22]. The level of H_2S in H_2 required to form bulk PtS is 20 times higher than that needed to form FeS at 700°K. Similar differences in the stability of the surface sulfides also exist (Section 6.2). An extreme example is copper, where the level of H_2S required to form a bulk sulfide is calculated to be 73%, one thousand times as high as that required in the case of Fe. The application of controlled sulfidation to bimetallic systems further extends the possibilities for controlling catalytic properties.

The use of zeolites as supports to impart sulfur resistance to metals through metal–zeolite interaction is receiving attention. Increased resistance to sulfur poisoning has been reported for isomerization and hydrogenation reactions [23–25]. In some of this work, the sulfur level was quite high, up to 0.5% H_2S by volume. The effect is attributed to an electron transfer from the metal crystallites within the supercage to the zeolite lattice. This may also affect the selectivity during hydrocarbon synthesis.

Supported metal alloys and bimetallic and multimetallic clusters (Chapter 2) represent another area which may impact on sulfur-resistant catalysts. While much work currently in progress is concerned with the effects of alloying on selectivity and activity, only in a few instances was sulfur resistance specifically studied. Little is known of the behavior of such systems in sulfur-free or sulfur-containing systems, but the mechanism of sulfur poisoning at levels of 1–100 ppm H_2S is probably a competition with the reactants for adsorption sites (Section 6.2). The formation of a surface sulfide may be strongly affected by addition of a second component to the catalytic metal. Strong compound formation may inhibit surface sulfidation. Alternatively a strongly electron-withdrawing second component may make the active metal more sulfur resistant, as is postulated in the case of Pt/zeolite catalysts [26]. Activity and selectivity data for copper alloyed with other active metals such as Ni would be of great interest as a result of the small tendency of Cu to form a bulk sulfide.

19.3.2 Long-Term Programs

19.3.2.1 New Catalytic Materials

The chance of finding unusual catalysts among the new materials available from recent research in inorganic and organometallic chemistry, materials science, and surface science makes such an evaluation worthwhile as a long-term program. The large number of possible materials that could be tested make a systematic approach very important. Ideally, a firm understanding of the mechanism of hydrocarbon synthesis would be of great help in such a test program, but the absence of information in this area, even on very simple systems (Section 7.1.2), makes mechanism studies an important aspect of supporting research.

The desired catalytic properties would preferably relate to selectivity and poison resistance. Activity would not be a primary concern initially. Considering selectivity first, the major problem with current processes is the lack of control on the product distribution. As the data presented in Table 19-1 show, product distributions achieved in hydrocarbon synthesis tend to be broad with a pronounced peak at methane. The C_{2+} hydrocarbons have a narrow maximum at C_4 to C_6 or a broad maximum extending from C_3 to C_{16}. The hydrocarbons produced are mostly normal parafins or α olefins and are not very useful as motor fuels because of their low octane quality. The production of high-octane hydrocarbon materials such as isoparaffins or aromatics is thermodynamically possible but there is no specific driving force for their synthesis. Catalyst selectivity, i.e., kinetics, is the controlling factor in the production of specific chemicals, among the many which are thermodynamically accessible.

Hydrocarbon synthesis catalysts require both hydrogenation activity and polymerization or CO insertion activity necessary for chain building. Alloy formation or multimetallic cluster formation is a promising route to alter the catalytic properties of a metal surface. Some systematic understanding of the surface properties of alloys is emerging from recent work on a variety of metals and reactions. The detailed evaluation of the application of alloy systems to hydrocarbon synthesis is a long-term research effort. The approach taken by Sinfelt and co-workers in the development of petroleum reforming catalysts is a good example [19]. This group began with the determination of the specific activity of a broad range of supported metals for several model reactions as a first step towards the understanding of the metal properties important for the reactions studied. A self-consistent model relating the reaction kinetics from one metal to another also resulted from this work. Then combinations of metals were tested where the choice of each component was made to either enhance or inhibit one part of the catalytic process.

In the case of hydrocarbon synthesis, Vannice has measured the specific activities of all the Group VIII metals at atmospheric pressure and several H_2/CO ratios [18]. This study should be extended to H_2/CO ratios below 2, to higher pressures to check the effect on product distribution, to other metals, especially to those of Group IB, and to well-characterized alloy systems.

Preliminary work on several alloy systems shows a definite effect of composition on selectivity. Some of these investigations were directed primarily at methanation, utilizing high H_2/CO ratios, and therefore are of limited use in the present context, while others are directed toward C_{2+} hydrocarbon synthesis. Bartholomew recently reported preliminary results comparing supported Ni and Ni combined with Mo, Co, Pt, Ru, Rh, and Pd [27]. Since the noble metals have a much higher selectivity for methane, it is not surprising that Ni–Pt/Al_2O_3 catalysts show very high selectivity for CH_4 while addition of Co or Ru depresses slightly the production of CH_4. The effects of alloying may be enhanced by the use of conditions more appropriate to the synthesis of higher hydrocarbons. An investigation of Pt–Fe alloys using Mössbauer spectroscopy to characterize the state of the iron shows that Fe alloyed with Pt loses its activity for synthesis and shifts the product distribution to lower molecular weight [28]. This effect has been attributed to a shift in electron density from the Fe to the Pt.

Another very interesting alloy system is that recently disclosed by Union Carbide [29] in which silica-gel-supported Rh–Fe catalysts were employed to make C_2 oxygenated compounds at enhanced efficiency. As the results of this patent collected in Table 19-3 show, the addition of Fe to Rh increases the production of ethanol at 1000 psig, and that of the sum of acetaldehyde and ethanol at 2500 psig at the expense of acetic acid production. In most of these data, the methane level is still quite high but the selective production of C_2 oxygenated species is encouraging and additional catalyst and process development may increase the selectivity. Metal dispersions (i.e., fraction of metal at the surface) reported for some of the catalysts used in the above work ranged from 2.7 to 22%. These values can be significantly increased with a possibly large effect on catalyst performance. In order to accomplish high dispersion of alloys on supports, it may be necessary to resort to the use of polynuclear organometallic complexes, thus minimizing the amount of unalloyed components present that cause poor selectivity.

Bimetallic catalysts may also find applications in preventing carbon deposition and increasing thermal stability. The inhibition of coke formation could result by preventing the formation of contiguous carbon networks on the metal surface or from an adventitious balance of hydrogenation and hydrogenolysis activity [19]. Thermal stability is only of secondary importance until a highly selective catalyst is obtained.

Another use for bimetallic and other composite materials would be as sulfur-resistant catalysts. It was shown (Section 17.2) that sulfidation at low levels of H_2S can be inhibited by combining the catalytic materials with a second component with which it forms a chemical bond. Since a surface sulfide layer forms even at very small ratios of H_2S/H_2 [30], the required heat of formation may be quite high. Formation of a simple alloy, for example NiCu where the bulk CuS phase is much less stable than NiS, may be inadequate. Benard has shown that Cu forms a complete surface sulfide layer at 850°C at 100 ppm of H_2S in H_2 [30]. At temperatures of 200–300°C employed in hydrocarbon synthesis, H_2S levels below 1 ppm may be sufficient to form a sulfide monolayer.

Borides, phosphides, and similar materials have higher heats of formation and may resist surface sulfidation. They are more resistant to sulfide formation than many metals, but their activity in Fischer–Tropsch is unknown. Similarly, intermetallic compounds or Brewer compounds of the type $ZrPt_3$ and $ZrIr_3$ have high heats of formation which impart high chemical and thermal stability (Section 10.2) and perhaps unusual catalytic properties worth studying.

In view of the experience in reforming, it should be feasible to use sulfur to control product selectivity in the Fischer–Tropsch synthesis. Sulfur addition is used industrially to modify the excessive activity of fresh naphtha reforming catalysts and also to improve product selectivity [31]. Recent work has shown that continuous additions of sulfur to operating Pt reforming catalysts results in improved selectivity and catalyst life [32,33]. The proposed mechanism is a partial poisoning of the Pt surface by sulfur to reduce the concentration of

TABLE 19-3

SYNTHESIS OF HYDROCARBONS ON SUPPORTED RhFe CATALYSTS[a]

Temperature (°C)	Pressure (psig)	GHSV (h^{-1})	Catalyst composition		C efficiency (%)				
			%Rh	%Fe	Methane	Methanol	Acetaldehyde	Ethanol	Acetic acid
300	1000	4000	5.0	—	37	0	24.0	16	20.0
300	1000	3900	—	1.0	69	12	0.3	10	0.2
300	1000	2700	2.5	1.35	51	19	0.6	23	0.6
300	1000	3000	2.5	0.68	44	20	1.0	30	1.3
325	2500	11000	2.5	—	41.0	1.6	5.2		49.0
325	2500	9500	2.5	1.35	39.0	13.0	37.0		4.0
325	2500	9300	2.5	0.135	5.2	2.3	13.0		31.0
325	2500	8700	2.5	0.135	46.0	1.4	15.0		36.0

[a] From Bhasin [29].

multiple surface sites that can result in excessive dehydrogenation and coking. The concentrations of S tested were quite high, up to 2400 ppm, and even at these levels good catalyst activity was maintained.

In Fischer–Tropsch synthesis, low S levels have been found to cause an increase in liquid hydrocarbons while a larger amount shifted the product distribution back to gaseous hydrocarbons. Encouraging results were obtained on supported Ni and Ru catalysts where methane formation was the primary interest [20]. At 400°C, an H_2/CO ratio of 4, and at atmospheric pressure, the product was between 90 and 100% methane. Addition of 1 ppm H_2S to the reactant stream and subsequent saturation of the catalyst gave a steady-state product selectivity of approximately 50% methane and 50% C_{2+} hydrocarbons. As the H_2S level was increased to 10 ppm, the products shifted further toward C_{2+} hydrocarbons. At lower temperatures and conditions more representative of liquid hydrocarbon production, similar effects of sulfur are expected.

19.3.2.2 Bifunctional Catalysts

Since with the exception of methane no particular type of hydrocarbon has a significantly high free energy of formation, the problem of hydrocarbon synthesis becomes one of catalytic selectivity. One class of products thermodynamically preferred is that of the isoparaffins. The large proportion of normal paraffins found in the products is a result of kinetic effects on these catalytic materials. Apparently isoparaffins and cyclic material must be generated by secondary reactions of the straight-chain hydrocarbons.

The proportion of isoparaffins in the product stream can, in principle, be increased by including an acidic isomerization component in the catalyst. Thus, the catalyst may function as a bifunctional catalyst with the normal paraffins generated on the metal component and subsequently isomerized on an acidic oxide support. A problem may exist in matching the reaction conditions for each of these processes. Synthesis is currently carried out at temperatures of 180–320°C in relatively small excess of hydrogen. The latter is necessary to minimize the deposition of coke on and the degradation of the acidic oxide surface. However, isomerization systems may exist in which the conditions are comparable and bifunctional systems may be practical. As an example, the Penex Process of Universal Oil Products operates at 1500–2000 psig and 120–150°C and isomerizes C_4 to C_6 paraffins [34]. The particular catalyst used is formed by reacting alumina-supported platinum with organic chlorides.

The production of aromatic hydrocarbons by the dehydrocyclization of C_{6+} hydrocarbons would be an additional benefit connected with using bifunctional catalysts since aromatics are high-octane constituents of gasoline.

Recently ERDA announced the funding of a cooperative research effort with Mobil Oil Corporation for the further development of a single-step process for the conversion of CO and H_2 directly to high-octane gasoline using bifunctional zeolite catalysts [35].

A variation of this process [36–41] is a good example for overcoming a selectivity problem by designing a reaction around products which have high stability or are accessible via a selective catalytic step. In this process, H_2 and CO are reacted first over a catalyst composed of a hydrogenation component and an acidic dehydration component to a product composed mainly of dimethyl ether. The hydrogenation catalyst is similar to current low-temperature methanol synthesis catalysts. The inclusion of an acid dehydration catalyst converts the methanol to dimethyl ether and in this manner shifts the reaction equilibrium to significantly higher conversions. The products are then directed over a second catalyst of crystalline aluminosilicate with high silica to alumina ratio and small pore diameters to produce liquid hydrocarbons with high aromatic content. The important aspect of this approach is that the desired intermediate product, dimethyl ether, can be produced in high yields on a bifunctional catalyst of high selectivity. Although the dimethyl ether is not a desired product, it serves a useful function by being both a readily available synthesis product and also a reactant that can be selectively converted to aromatic hydrocarbons. The acid dehydration and shape-selective properties of the aluminosilicate catalyst must be carefully controlled to give high aromatic yield.

19.3.2.3 Isosynthesis

Isosynthesis was first developed in the early 1940s and has received very little attention since. The early work has been reviewed by Cohn [42]. In its original form, a catalyst consisting of thoria promoted by alumina and alkali is operated at 450°C and 300–600 atm, with a H_2/CO ratio of 2:1. The product consists of isoparaffins, mainly isobutane, some branched olefins, and aromatics. The product distribution is centered at low molecular weights, C_1 to C_4 hydrocarbons representing over 50% of the product. The ability to produce branched and aromatic hydrocarbons makes the isosynthesis an important process for the production of high-octane motor fuel. The product distribution must be shifted to higher carbon numbers and the activity and reaction conditions improved if the process is to become commercially viable.

Alternatively, the process may be optimized to selectively produce isobutane or olefins. Some of the better catalysts were capable of producing isobutane at 60% selectivity, while operation at high pressure favored dimethyl ether productions.

19.3.2.4 Homogeneous Hydrocarbon Synthesis

Such synthesis from CO and H_2 with mononuclear catalysts has not yet been accomplished. Muetterties reported the synthesis of hydrocarbons on the homogeneous cluster complexes, $Os_3(CO)_{12}$ and $Ir_4(CO)_{12}$ [43]. At 140°C and with a CO and H_2 pressure of 2 atm, a turnover number of 10^{-5} sec^{-1} per cluster was observed, a rate comparable to that observed by Vannice on silica-supported Ir [21]. The reaction showed a high selectivity to methane. Substitution of carbonyl groups in $Ir_4(CO)_{12}$ by triphenyl phosphine caused a substantial shift in the product selectivity to ethane and propane, thus providing an example for changing the product distribution on homogeneous cluster complexes.

Another type of synthesis over homogeneous catalysts is a process recently patented by Union Carbide Corporation. In one version, a mononuclear rhodium complex such as 2-pyridino-1-dicarbonyl(2,4-pentanedionato)rhodium is reacted with H_2 and CO ($H_2/CO = 1.5$) at 20,000 psi to produce ethylene glycol and methanol [44]. The need for high pressures is a disadvantage, but recent improvements allow operation at lower pressures. The use of rhodium dicarbonylacetylacetonate-triisopropanolamine, or cluster complexes such as $Rh_6(CO)_{15}$ in the presence of Cs formate, ammonium acetate, or triphenylsilane as the promoter requires a CO and H_2 pressure of 8000 psi to give reasonable yields of ethylene glycol and methanol [45]. If process conditions can be improved, direct glycol manufacture from CO and H_2 can have economic advantages [46].

Another homogeneous process that utilizes synthesis gas to produce hydrocarbons is homologation in which methanol is reacted with CO and H_2 to produce higher alcohols, generally ethanol. The catalyst employed in the initial work was dicobalt octacarbonyl $Co_2(CO)_8$ and the conditions were quite stringent, 5100 psi H_2/CO (1:1) and 185°C [47,48]. Ethanol represented over 50% of the product. More recent work investigated the effect of promoters such as iodine, organophosphorus compounds [49], or iodides [50]. In the latter case at 200°C and 4500–6000 psi the ethanol yield was 40% at a methanol conversion of 60%.

In both the homologation chemistry and in the Union Carbide process described earlier, the products are oxygenated hydrocarbons. Current understanding of the mechanisms of homogeneous catalysis support the concept that the scission of the carbon–oxygen bond cannot be accomplished by the single metal center of a mononuclear homogeneous complex [43]. This is consistent with the production of ethylene glycol by the Union Carbide mononuclear rhodium catalyst where the C–O bonds remain intact. However, the homologation adds a CH_2 to the methanol in a step similar to the chemistry of Fischer–Tropsch synthesis. This includes C–O bond breaking and as such would require a multinuclear complex to be consistent with the concept outlined above. This particular approach to homogeneous CO–H_2 synthesis requires more investigation, especially as it relates to the search for homogeneous catalysts for hydrocarbon synthesis.

19.3.3 Supporting Research

Basic understanding of the mechanism of catalysis and the surface properties important in controlling activity and selectivity of CO hydrogenation would be extremely useful in the development of catalysts and in the search for new catalytic materials. Research in two areas is needed. First, the mechanism of formation of C–C bonds on such common catalysts as reduced Fe and Co should be established. The mechanism involving the polymerization of a surface species and competition between polymerization and termination to control hydrocarbon chain length is somewhat speculative and based on indirect evidence. In both methanation and Fischer–Tropsch synthesis a partially hydrogenated surface enol of the form HCOH was postulated and its reaction to methane or condensation to form C–C hydrocarbon bonds was assumed to be the slow step. Recent data, however, indicate that the slow step may be scission of the C–O bond of the adsorbed CO. A number of recent experimental results support this particular mechanism. The measurement of the kinetic isotope effect showed that, on supported Ni, Ru, and Pt, the reactions of H_2 + CO and of D_2 + CO proceed at identical rates, implying that hydrogen is not involved in the rate-determining step [51]. Studies on Ni and Ni–Cu alloys show that an ensemble of contiguous surface sites is required to dissociate C–O prior to reaction with hydrogen [52]. According to recent LEED and UPS measurements on Ni, co-adsorption of H_2 and CO does not lead to formation of a surface enol complex and decomposition of the CO may be required before hydrogenation of the CO can proceed [53]. These data are consistent with in situ infrared spectroscopic examination of alumina-supported Ru, Rh, and Pt catalysts, indicating that the surface is covered predominately by adsorbed CO during reaction of H_2 and CO without any evidence of a surface formyl complex [54]. Important surface properties such as strength of CO binding, simultaneous adsorption of CO and H_2, and the need for other structural or electronic properties should be clarified. This should aid in understanding the selectivity variation observed between metals and the action of promoters such as potassium.

The second line of basic research related to that described above is the comparative study of H_2–CO reaction on a variety of materials that show widely different selectivities. Some materials, such as ZnO and Cu, selectively produce methanol. Iron and Co produce alcohols, olefins, and paraffins under certain conditions and almost exclusively paraffins under other conditions. Still other metals, Ni and Pt for example, produce mostly methane. Such variation is generally ascribed to relative hydrogenation activity of the catalytic material, but the controlling factors in such gross selectivity changes are not well understood. With such widely varying surfaces and metal oxidation states, the adsorbed species or reaction intermediates may well be different. Certainly the stability and binding strength of a surface species would be quite different on such a variety of materials.

19.4 CONCLUSIONS

The product distribution in hydrocarbon synthesis from CO and H_2 is kinetically controlled in nearly all cases. The greater stability of methane leads to significant yields for a variety of conditions and catalysts, a major problem when liquid hydrocarbons are the desired products. The difficulty in obtaining a kinetic selectivity arises from the mechanism of hydrocarbon synthesis, a polymerization process. However, past catalyst performance teaches that the position of the product distribution can be shifted and narrowed somewhat by changes in operating conditions or in catalyst composition. The desire for even more specific processes makes further development of synthesis catalysts imperative. Similarly, the development of sulfur-resistant catalysts capable of operating at 10–1000 ppm levels of sulfur would greatly improve the efficiency of the hydrocarbon synthesis process.

The effects of promoters, supports, and preparation procedures on catalyst activity and selectivity represent good subjects for short-term development studies. The preparation of highly dispersed supported catalysts and multicomponent materials by novel preparative techniques will permit an extensive evaluation of support and promoter effects. However, it should be borne in mind that the test conditions are very important and can greatly affect the activity and product distribution. The tests should be carried out not only under conditions used in the current process but also at varied temperatures, pressures, and H_2/CO ratios with particular attention to product analysis and to changes in catalyst selectivity.

Alloys, complex oxides, and new catalytic materials should be explored because of their potential interest as catalysts for hydrocarbon synthesis. They should also be tested for sulfur resistance. Studies on well-characterized alloy systems such as NiCu, CuRu, PdAu, and cluster oxides, as Mg_2MoO_8, represent short-term efforts since some information on their catalytic activity for related reactions is available. The development of optimum alloys systems and the clarification of important catalytic properties, such as oxidation state, electronic, and structural properties, will be of a long-term nature and will be greatly aided by in-depth studies of the mechanism of higher hydrocarbon synthesis. A better understanding of the surface properties controlling the course of the synthesis reaction represents another important basic research objective where effort is needed.

Finally, bifunctional catalysts and multistep processes have recently been shown to be extremely promising areas. New chemistry exemplified by the processes discovered by Mobil should be further developed and extended. The multistep processing of $CO–H_2$ mixtures may permit the selective production of a wide variety of hydrocarbon materials.

REFERENCES

1. Anderson, R. B., *in* "Catalysis" (P. H. Emmett, ed.), Vol. IV, p. 1. Van Nostrand–Reinhold, Princeton, New Jersey, 1956.
2. Storch, H. H., Golumbic, N., and Anderson, R. B., "The Fischer–Tropsch and Related Syntheses." Wiley, New York, 1951.
3. Friedel, R. A., and Anderson, R. B., *J. Am. Chem. Soc.* **72**, 1212, 2307 (1950).
4. Weitkamp, A. W., Seelig, H. S., Bowman, W. J., and Cady, W. E., *Ind. Eng. Chem.* **45**, 343 (1953).
5. Pichler, H., and Schulz, H., *Chem. Eng. Tech.* **42**, 1162 (1970).
6. Nefedov, B. K., and Eidus, Ya T., *Russ. Chem. Rev.* **34**, 272 (1965).
7. Bienstock, D., Jimeson, R. M., Field, J. H., and Benson, H. E., U.S. Bureau of Mines R. I. 5655 (1960); Field, J. H., Bienstock, D., Forney, A. J., and Demski, R. J., U.S. Bureau of Mines R. I. 5871 (1961).
8. Karn, F. S., Schultz, J. F., Kelly, R. E., and Anderson, R. B., *Ind. Eng. Chem. Prod. Des. Dev.* **2**, 43 (1963).
9. Karn, F. S., Shultz, J. F., Kelly, R. E., and Anderson, R. B., *Ind. Eng. Chem. Prod. Des. Dev.* **3**, 33 (1964).
10. Herington, E. F. G., and Woodward, L. A., *Trans. Faraday Soc.* **35**, 958 (1939).
11. Craxford, S. R., *Trans. Faraday Soc.* **42**, 576 (1946).
12. White, G. A., Roszkowski, T. R., and Stanbridge, D. W., *Am. Chem. Soc. Div. Fuel Chem. Prepr.* **19** (3), 57 (1974); L. Seglin, (ed.), "Methanation of Synthesis Gas," p. 138. American Chemical Society, Washington, D.C. 1975.
13. Greyson, M., *in* "Catalysis" (P. H. Emmett, ed.), Vol. IV, p. 473. Van Nostrand–Reinhold, Princeton, New Jersey, 1956.
14. Field, J. H., Bienstock, D., Forney, A. J., and Demski, R. J., U.S. Bureau of Mines R. I. 5871 (1961).
15. Hoogendoorn, J. C., presented at *Clean Fuels Coal Symp.* (1973).
16. *Chem. Eng. News,* p. 24, July 19, 1976.
17. Vannice, M. A., and Garten, R. L., U.S. Patent No. 3,941,819, March 2, 1976.
18. Vannice, M. A., *J. Catal.* **37**, 449 (1975).
19. Sinfelt, J. H., Carter, J. L., and Yates, D. J. C., *J. Catal.* **24**, 283 (1972); Sinfelt, J. H., Barnett, A. E., and Carter, J. L., U.S. Patent No. 3,617,518 (1971).
20. Dalla Betta, R. A., Piken, A. G., and Shelef, M., *J. Catal.* **40**, 173 (1975).
21. Vannice, M. A., *J. Catal.* **37**, 462 (1975).
22. Boudart, M., Cusumano, J. A., and Levy, R. B., New Catalytic Materials for the Liquefaction of Coal, Rep. RP-415-1, Electric Power Res. Inst., October 30, 1975.
23. Dalla Betta, R. A., and Boudart, M., *in* "Catalysis" (J. Hightower, ed.), p. 1329. North-Holland Publ., Amsterdam, 1972; Gallezot, T., Alarcon-Diaz, A., Dalmon, J. A., Renouprez, J. R., and Imelik, B., *J. Catal.* **39**, 334 (1975).

24. Gallezot, P., Datka, J., Massardier, J., Primet, M., and Imelik, B., *Proc. Int. Congr. Catal. 6th* (G. C. Bond *et al.*, eds.), p. 696. The Chemical Society, London, 1977.
25. Chukin, G. D. *et al.*, *Proc. Int. Congr. Catal. 6th* (G. C. Bond *et al.*, eds.), p. 668. The Chemical Society, London, 1977.
26. Rabo, J. A., Schomaker, V., and Pickert, P. E., *Proc. Int. Congr. Catal., 3rd, 1964*. Vol. 2, p. 1264. North-Holland Publ., Amsterdam, 1965.
27. Bartholomew, C. H., ERDA Project No. E(49-18)-1790, Annual Technical Progress Rep., May 6, 1976.
28. Vannice, M. A., and Garten, R. L., *J. Mol. Catal.* **1**, 201 (1976).
29. Bhasin, M. M., German Patent No. 2,503,233, July 31, 1975.
30. Benard, J., *Catal. Rev.* **3** (1), 93 (1969).
31. Maxted, E. B., *Adv. Catal. Relat. Sub.* **3**, 129 (1951).
32. Haensel, V., Donaldson, G. R., Riedl, F. J., *Proc. Int. Congr. Catal., 3rd, Amsterdam* p. 298 (1964).
33. Hayes, J. C., Mitsche, R. T., Pollitzer, E. L., and Homeier, E. H., *Am. Chem. Soc. Div. Pet. Chem. Prepr.* **19** (2), 334 (1974).
34. Thomas, C. L., "Catalytic Processes and Proven Catalysts," p. 17. Academic Press, New York, 1970.
35. Heinemann, H., private communication, 1975.
36. Chang, C. D., and Silvestri, A. J., U.S. Patent No. 3,894,102 (July 8, 1975).
37. Chang, C. D., Silvestri, A. J., and Smith, R. L., U.S. Patent No. 3,894,103 (July 8, 1975).
38. Chang, C. D., Lang, W. H., and Silvestri, A. J., U.S. Patent No. 3,894,104 (July 8, 1975).
39. Chang, C. D., Silvestri, A. J., and Smith, R. L., U.S. Patent No. 3,894,105 (July 8, 1975).
40. Chang, C. D., Lang, W. H., and Silvestri, A. J., U.S. Patent No. 3,894,106 (July 8, 1975).
41. Butler, S. A., Jurewicz, and Kaeding, W. W., U.S. Patent No. 3,894,107 (July 8, 1975).
42. Cohn, E. M., *in* "Catalysis" (P. H. Emmett, ed.), Vol. IV, p. 443. Van Nostrand–Reinhold, Princeton, New Jersey, 1976.
43. Thomas, M. G., Beier, B. F., and Muetterties, E. L., *J. Am. Chem. Soc.* **98**, 1296 (1976).
44. Walker, W. E., and Cropley, J. B., German Patent No. 2,426,495 (June 19, 1973); *Chem. Abstr.* **82**, 139312Q; Pruett, R. L., and Walker, W. E., U.S. Patent No. 3,833,634 (September 3, 1974).
45. Kaplan, L., U.S. Patent No. 3,944,588 (January 2, 1976); *Chem. Abstr.* **84**, 164152M.
46. Brownstein, A. M., *Chem. Eng. Progr.* **71**, 72 (1975).
47. Wender, I., Friedel, R. A., and Orchin, M., *J. Am. Chem. Soc.* **71**, 4160 (1949).
48. Wender, I., Friedel, R. A., and Orchin, M., *Science* **113**, 206 (1951).
49. U.S. Patent No. 3,248,432 (April 26, 1966).
50. Mizoraki, T., and Nakayama, *Bull. Chem. Soc. Jpn.* **37**, 236 (1964).
51. Dalla Betta, R. A., and Shelef, M., *J. Catal.* **49,** 383 (1977).
52. Araki, M., and Ponec, V., *J. Catal.* **44,** 439 (1976).
53. Conrad, H., Ertl, G., Küppers, J., and Latta, E. E., *Proc. Int. Congr. Cataly., 6th* (G. C. Bond *et al.*, eds.), p. 427. The Chemical Society, London, 1977.
54. Dalla Betta, R. A., and Shelef, M., *J. Catal.* **48,** 111 (1977).

INDEX

P

Q

R

S

A
B
C 8
D 9
E 0
F 1
G 2
H 3
I 4
J 5